SpringerBriefs in Petroleum Geoscience & Engineering

Series Editors

Dorrik Stow, Institute of Petroleum Engineering, Heriot-Watt University, Edinburgh, UK

Mark Bentley, AGR TRACS International Ltd, Aberdeen, UK

Jebraeel Gholinezhad, School of Engineering, University of Portsmouth, Portsmouth, UK

Lateef Akanji, Petroleum Engineering, University of Aberdeen, Aberdeen, UK

Khalik Mohamad Sabil, School of Energy, Geoscience, Infrastructure and Society, Heriot-Watt University, Edinburgh, UK

Susan Agar, Oil & Energy, Aramco Research Center, Houston, USA

Kenichi Soga, Department of Civil and Environmental Engineering, University of California, Berkeley, USA

A. A. Sulaimon, Department of Petroleum Engineering, Universiti Teknologi PETRONAS, Seri Iskandar, Malaysia

The SpringerBriefs series in Petroleum Geoscience & Engineering promotes and expedites the dissemination of substantive new research results, state-of-the-art subject reviews and tutorial overviews in the field of petroleum exploration, petroleum engineering and production technology. The subject focus is on upstream exploration and production, subsurface geoscience and engineering. These concise summaries (50–125 pages) will include cutting-edge research, analytical methods, advanced modelling techniques and practical applications. Coverage will extend to all theoretical and applied aspects of the field, including traditional drilling, shale-gas fracking, deepwater sedimentology, seismic exploration, pore-flow modelling and petroleum economics. Topics include but are not limited to:

- Petroleum Geology & Geophysics
- Exploration: Conventional and Unconventional
- Seismic Interpretation
- Formation Evaluation (well logging)
- Drilling and Completion
- Hydraulic Fracturing
- Geomechanics
- Reservoir Simulation and Modelling
- Flow in Porous Media: from nano- to field-scale
- Reservoir Engineering
- Production Engineering
- Well Engineering; Design, Decommissioning and Abandonment
- Petroleum Systems; Instrumentation and Control
- Flow Assurance, Mineral Scale & Hydrates
- Reservoir and Well Intervention
- Reservoir Stimulation
- Oilfield Chemistry
- Risk and Uncertainty
- Petroleum Economics and Energy Policy

Contributions to the series can be made by submitting a proposal to the responsible Springer contact, Charlotte Cross at charlotte.cross@springer.com or the Academic Series Editor, Prof Dorrik Stow at dorrik.stow@pet.hw.ac.uk.

More information about this series at http://www.springer.com/series/15391

Jiefu Chen · Shubin Zeng ·
Yueqin Huang

Borehole Electromagnetic Telemetry System

Theory, Modeling, and Applications

Prof. Jiefu Chen
Department of Electrical and Computer
Engineering
University of Houston
Houston, TX, USA

Shubin Zeng
Department of Electrical and Computer
Engineering
University of Houston
Houston, TX, USA

Dr. Yueqin Huang
Cyentech Consulting LLC
Cypress, TX, USA

ISSN 2509-3126   ISSN 2509-3134 (electronic)
SpringerBriefs in Petroleum Geoscience & Engineering
ISBN 978-3-030-21536-1   ISBN 978-3-030-21537-8 (eBook)
https://doi.org/10.1007/978-3-030-21537-8

This Springer imprint is published by the registered company Springer Nature Switzerland AG
The registered company address is: Gewerbestrasse 11, 6330 Cham, Switzerland

Contents

Contents

Acronyms

1D	One dimensional
2D	Two dimensional
3D	Three dimentional
BOP	Blowout preventer
CSEM	Controlled source electromagnetic
EFIE	Electric field integral equation
EGS	Enhanced geothermal system
EMT	Electromagnetic telemetry
ERT	Electrical resistivity tomography
FEM	Finite element method
GL	Gaussian–Legendre
IE	Integral equation
LMGF	Layered media Green's function
LWD	Logging-While-Drilling
MPT	Mud pulse telemetry
MT	Magnetotellurics
MWD	Measurement-While-Drilling
NMM	Numerical mode matching
OBM	Oil-based mud
SI	Sommerfeld integral
TE	Transverse electric
TI	Transversely isotropic
TM	Transverse magnetic
TVD	True vertical depth
VEP	Volumetric equivalence principle
WAM	Weighted average method
WBM	Water-based mud

Chapter 1
Introduction

1.1 Oilfield Drilling and Downhole Wireless Communication

Petroleum and natural gas are among the primary energy sources for the United States and around the world. According to "International Energy Outlook 2017" released by the Energy Information Administration (EIA), 66% of U.S. energy consumption was from petroleum and gas in 2016 [4]. Looking forward to 2040, more than 60% of energy consumption still depends on petroleum and natural gas. In the years to come, significant efforts on exploration and production (E&P) of subsurface resources are needed to keep up with the global energy consumption demands. Although there are still many oil and gas reserves left to be discovered and produced, many of them are in areas difficult to access (e.g. deepwater and ultra-deep water), or they are in the form of unconventional resources such as oil shale and shale gas.

To tap the difficult oil and natural gas deposits, the oil and gas industry is striving to develop new technologies to squeeze more fuel from existing reserves and reach previously inaccessible fields. Directional and horizontal drilling can adjust well trajectory to maximize reservoir exposure and increase well productivity. For example, in offshore oil and gas exploration shown in Fig. 1.1, directional drilling and horizontal drilling substantially reduces the costs by reaching multiple geological targets from a single platform. In the case of hydraulic fracturing shown in Fig. 1.2, the horizontal section of a drilled wellbore, which can be thousands of feet long, needs to be kept in the desired shale formation. The drilling direction as well as the wellbore placement with respect to shale formation boundaries will greatly affect the fracturing efficiency and shale gas production. While drilling, a variety of measurements, such as temperature, pressure, resistivity logging, nuclear logging, etc. are acquired in the borehole environments, and then sent to the surface to ensure safety, to adjust the well trajectory on the fly to reach the geological targets, and to keep the wellbore within a desired formation. The process of sending downhole measurements to the surface, or sending drilling commands from the surface to the downhole tool, is

J. Chen et al., *Borehole Electromagnetic Telemetry System*,
SpringerBriefs in Petroleum Geoscience & Engineering,
https://doi.org/10.1007/978-3-030-21537-8_1

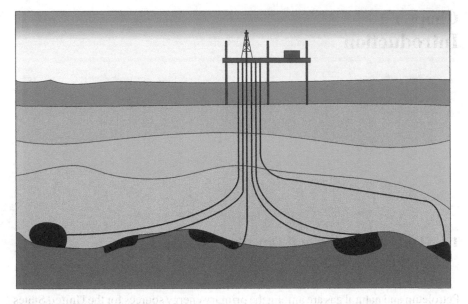

Fig. 1.1 Directional drilling for reaching multiple geologic targets from a single platform in off-shore exploration

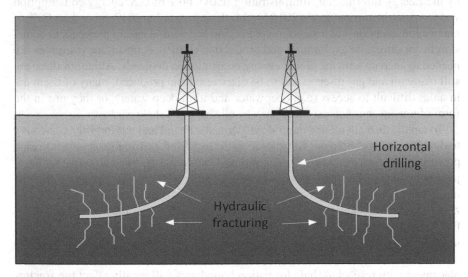

Fig. 1.2 Horizontal drilling for keeping the wellbore within the desired shale formation in shale gas hydraulic fracturing

called telemetry. Because no cable can be deployed in the borehole during drilling, borehole telemetry in oilfield drilling is achieved by wireless communication.

1.2   Electromagnetic Telemetry

There exist two widely used borehole wireless communication technologies: mud pulse (MP) and electromagnetic (EM) telemetry. As shown in Fig. 1.3, mud pulse telemetry (MPT) employs a downhole valve to restrict the flow of drilling mud and to create pressure pulses carrying digital information and propagating along the borehole. MPT made possible real-time measurement-while-drilling (MWD), with the first commercial MWD system developed in the late 1970s. MPT is a mature and extensively adopted downhole communication technique, but it has severe limitations: it involves moving parts prone to failure during operation, and it cannot work if there is no continuous mud column in the borehole (when dust, mist, foam, or air used as fluid in underbalanced drilling), or when fluid loss of circulation is encountered (e.g. fractured rock that drains the drilling fluid from the borehole) [1].

Electromagnetic telemetry (EMT) is based on electromagnetic waves propagating in underground formation between downhole tool and the surface. As shown in Fig. 1.4, in an EMT system, a gap source near the drilling bit generates and propagates electromagnetic signals into the formations surrounding the drilled borehole, and the signals will be picked up on the surface by an earth antenna as a metal stake driven into ground with some distance from the rig. Compared with MPT, EMT has been proven to have better reliability as no moving part is involved during drilling, and it has the potential to provide a much higher rate of data transmission under certain circumstances [2]. Since EMT does not rely on the drilling fluid column as a communications channel, EMT systems are preferred in underbalanced drilling and in areas where loss of circulation is prevalent (e.g. shale gas or geothermal drilling).

In the past decade, a technical process termed "managed pressure drilling" has experienced favorable reception and adoption by the energy industry. The essential idea behind managed pressure drilling is to use as low a drilling mud density as is allowable under downhole circumstances, maintaining a balanced pressure on formations. Drilling a formation with balanced, or even slightly underbalanced pressure margin, results in faster rates of penetration. By exerting a very low differential pressure on the formations, the likelihood of experiencing pressure sticking of the drill pipe against the borehole wall is minimized. Significant efficiencies have been realized via the use of managed pressure, and underbalanced, drilling techniques. These techniques may at times utilize a very lightweight drilling fluid that may be injected with some concentration of nitrogen gas. In such circumstances, EMT systems are favored, as MPT systems will lose quality of telemetry in multi-phase fluids.

Following the 2014–2015 collapse in global oil prices, one of the chief means for operators to not only survive but maintain any reasonable profitability was to become more efficient at drilling wells. Reducing non-productive time (NPT) in drilling operations has become a significant factor in increasing efficiencies that ultimately

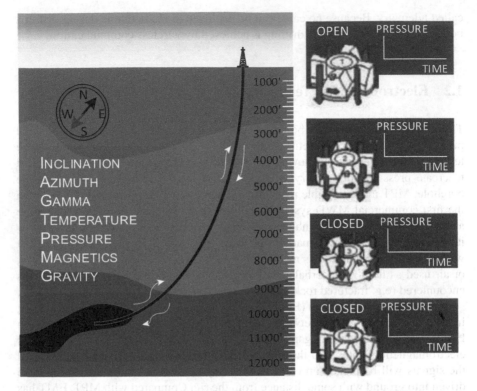

Fig. 1.3 A schematic of MPT: a downhole valve is operated to restrict the flow of drilling mud and to create pressure pulse carrying digital information and propagating along the borehole

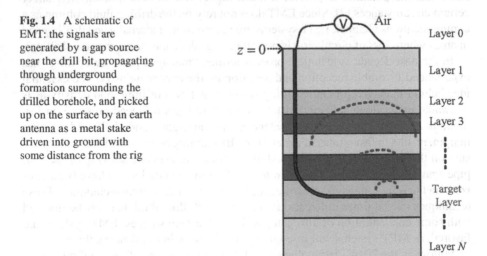

Fig. 1.4 A schematic of EMT: the signals are generated by a gap source near the drill bit, propagating through underground formation surrounding the drilled borehole, and picked up on the surface by an earth antenna as a metal stake driven into ground with some distance from the rig

lead to lower costs and higher profits. To this end, operators are embracing new and re-discovered technologies, such as EMT, that help reduce drilling costs while increasing their competitiveness. One of the operators' goals is to drill as complete an interval as possible in the least number of runs, to reduce the time and number of bit trips (thus to reduce NPT). With its telemetry independence from the borehole drilling fluid, complete EM-MWD directional surveys can be acquired and transmitted to the surface during a drill pipe connection, eliminating the 2–5 min of non-productive fluid circulation time required for a MP-MWD survey. With it not being unusual to have 200 MWD directional surveys on a 20,000-foot-deep well, NPT saving by EMT over 24 h on a well can be substantial. At a current U.S. land rig daily operating rate of $30,000, this cost reduction translates to significantly improved operator profitability. In summary, the direction the industry is heading, driven as it is to continually find greater efficiencies, increasingly favors EMT systems.

Despite all the admirable qualities and ensuing benefits of EMT systems, MPT remains the first choice in telemetry for drilling most wells. An EMT customer discovery report based on extensive interviews with oil and gas professionals suggest that lacking of fast and accurate EMT simulation software tools is one of the biggest challenges for EMT to penetrate the drilling market [6]. While the MPT signal strength is predominantly determined by drilling depth and can be easily predicted in a pre-job simulation, the received signal intensity of an EMT system is dependent upon a variety of factors, such as the structural and electromagnetic properties of underground formation (the earth model), borehole conditions, and deployment of surface antenna. Accurate and rapid prediction of EMT performance by rigorous numerical modeling is critical in justifying the employment of EMT over MPT for a specific job, and it is also essential in maximizing EMT received signals and optimizing for carrier frequency, bandwidth, and battery life as drilling depth increases.

Modeling an EMT system is essentially simulation of low frequency electromagnetic waves propagating through underground formation with borehole and drill string. One of the greatest challenges of EMT modeling is to discretize the complicated and multiscale EMT structure: the entire computational domain is on the order of thousands feet, and such a large volume will include hundreds of or even more formation layers; besides, the drilled borehole including drill string is very long (thousands of feet) and thin (several inches in diameter), and can contain even finer features (e.g. standoff between drill string and borehole wall is a fraction of an inch). Conventional numerical techniques such as the finite different method (FDM) or finite element method (FEM) will either fail to generate mesh for such an extremely multiscale structure, or lead to a huge number of unknowns (e.g. tens of millions) and prohibitively high computational costs.

Several EMT modeling and simulation methods have been developed and published, such as the work developed jointly by Geoservices and University of Lille [3], jointly by Shell and the University of Texas at Austin [9], by Norwegian University of Science and Technology [8], and by Colorado School of Mines [7]. But these models all possess limitations, such as only being applicable to vertical wells, requiring the drilling/casing string be perfect electric conductor (PEC), unable to deal with multiple-layer underground formation, incapable of handling formation anisotropy. None of the aforementioned methods can provide a rigorous and efficient (real-time) 3D EMT modeling service, which is indispensable for deviated and horizontal drilling. On the other hand, generic EM simulation software tools such as COMSOL and ANSYS are versatile in modeling different 3D EM telemetry systems in complicated environments, but usually take several hours to perform a simulation at just a single depth [5]. It would be unaffordably expensive to conduct a pre-job EMT study using commercial software as modeling an entire EMT job oftentimes requires rigorous EM simulations at (tens of) hundreds of drilling depths.

In the following chapters of this book, we will first discuss simulations of EMT in vertical wells, which are relatively simple 2D problems. Then we will discuss a more general 3D modeling technique based on integral equation method for EMT in directional and horizontal drilling. Finally, we will give various applications such as a long range EMT system using casing antenna and crosswell EMT in pad drilling, and discuss the performances of these systems in different situations based on rigorous analyses using the efficient numerical techniques discussed in this book.

References

1. Bennion, D.B., Thomas, F.B., Bietz, R.F., Bennion, D.W., et al.: Underbalanced drilling, praises and perils. In: Permian Basin Oil and Gas Recovery Conference. Society of Petroleum Engineers (1996)
2. Chen, J., Li, S., MacMillan, C., Cortes, G., Wood, D., et al.: Long range electromagnetic telemetry using an innovative casing antenna system. In: SPE Annual Technical Conference and Exhibition. Society of Petroleum Engineers (2015)
3. DeGauque, P., Grudzinski, R., et al.: Propagation of electromagnetic waves along a drillstring of finite conductivity. SPE Drill. Eng. 2(02), 127–134 (1987)
4. EIA, US: International Energy Outlook 2017. US Energy Information Administration (2017)
5. Jannin, G., Chen, J., DePavia, L.E., Sun, L., Schwartz, M.: Deep electrode: a game-changing technology for electromagnetic telemetry. In: SEG Technical Program Expanded Abstracts 2017, pp. 1059–1063. Society of Exploration Geophysicists (2017)
6. Shen, Q., Chen, J., MacMillan, C., Wu, X.: I-Corps: advanced 3D simulation software for electromagnetic telemetry. In: Final presentation of customer discovery. National Science Foundation I-Corps Project #1755200 (2017)
7. Tang, W., Li, Y., Swidinsky, A., Liu, J.: Three-dimensional controlled-source EM modeling with an energized well casing by finite element. In: International Workshop and Gravity, Electrical & Magnetic Methods and their Applications, pp. 394–397. Society of Exploration Geophysicists and and Chinese Geophysical Society (2015)
8. Wei, Y.: Propagation of electromagnetic signal along a metal well in an inhomogeneous medium. Ph.D. thesis, Norwegian University of Science and Technology (2013)
9. Yang, W., Torres-Verdn, C., Hou, J., Zhang, Z.I.: 1D subsurface electromagnetic fields excited by energized steel casing. Geophysics 74(4), E159–E180 (2009)

Chapter 2
2D Modeling of Electromagnetic Telemetry

2.1 Electromagnetic Telemetry in Vertical Drilling

In this chapter we will discuss the modeling of electromagnetic telemetry in vertical wells, whose schematic is shown in Fig. 2.1. A transmitter in this wireless communication system is near the drill bit implemented as a voltage gap source or a toroid, which can be viewed in numerical analysis as an electric dipole or as a magnetic current source circulating the drill string, respectively. A receiver is set on the surface with two terminals: one is attached to the blowout preventer (BOP), and the other one is connected to an earth antenna as a metal stake driven into the ground with a certain distance away from the rig. The performance of an electromagnetic telemetry system is strongly affected by properties of the underground formation, which oftentimes is treated as a layered structure along the vertical direction. On the other hand, the telemetry system in a vertical well can also be viewed as a cylindrically layered structure: several layers including steel drill string, drilling fluid in the borehole, one of multiple casing columns, one or multiple cement columns, and underground formation can be found from the center of wellbore to outside along the radial direction. Such a doubly layered structure with a gap source or a toroid source can be regarded as an axisymmetric and transverse magnetic (TM) problem.

Various methods have been developed for 2D modeling of EMT in vertical drilling, for example, the analytical method [1, 11, 15], the telegrapher method [2, 3], the 2D finite element method (FEM) [16, 17], the numerical mode matching (NMM) method [12, 13], and the integral equation method (IEM) [10]. Generally speaking, analytical or quasi-analytical methods are fast at the expense of flexibility in modeling complicated structures. Numerical methods such as FEM is more versatile and capable of taking all the features in an EMT system such as drill pipes, drilling fluid in the borehole, multiple layers of casing and cement, and earth formation into consideration. However, the computational costs of conventional FEM will grow dramatically when used to model a deep well, or when the number of underground layers is large. In this chapter we will introduce a recently developed semianalytical

© The Author(s), under exclusive license to Springer Nature Switzerland AG 2019
J. Chen et al., *Borehole Electromagnetic Telemetry System*,
SpringerBriefs in Petroleum Geoscience & Engineering,
https://doi.org/10.1007/978-3-030-21537-8_2

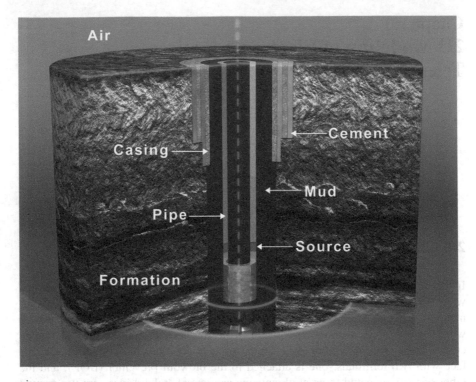

Fig. 2.1 A schematic of electromagnetic telemetry in vertical well

finite element method [5, 8, 9] for axisymmetric electromagnetic problems, and apply it for simulations of EMT systems in vertical wells.

2.2   Semianalytical FEM for Axisymmetric EMT Modeling

Using $e^{j\omega t}$ as the time convention, the governing equation for axisymmetric EMT modeling based on variable H_ϕ is

$$\frac{\partial}{\partial \rho}\left(\frac{1}{\rho \hat{\varepsilon}_r}\frac{\partial}{\partial \rho}\left(\rho H_\phi\right)\right) + \frac{\partial}{\partial z}\left(\frac{1}{\hat{\varepsilon}_r}\frac{\partial H_\phi}{\partial z}\right) + k_0^2 \mu_r H_\phi = j\frac{k_0}{Z_0}M_\phi \qquad (2.1)$$

where k_0 and Z_0 denote wavenumber and intrinsic impedance of the free space, respectively. μ_r is the relative permeability. Complex relative permittivity $\hat{\varepsilon}_r = \varepsilon_r - j\sigma/(\omega\varepsilon_0)$. ε_0 is permittivity of vacuum. ε_r and σ are relative permittivity and conductivity, respectively. There are two ways of implementing a downhole source: for voltage source, magnetic current M_ϕ circulating the drill string needs to be specified at the source position; for current source, the magnitude of current flowing along the pipe needs to be known, which is equivalent to the Dirichlet boundary condition for magnetic field at the source position. These two types of source implementation are equivalent to each other based on normalized output power, i.e., the multiplication of voltage jump at the source (known for the first implementation, and to be calculated for the second implementation) and current flow at the source (known for the second implementation, and to be calculated for the first implementation).

M_ϕ is imposed magnetic current density circulating the drill string. The functional corresponding to (2.1) is

$$
\Pi\left(H_\phi\right) = \pi \int_{z_a}^{z_b} \int_{\rho_a}^{\rho_b} \left[\frac{1}{\hat{\varepsilon}_r} \left(\frac{1}{\rho} \frac{\partial}{\partial \rho} \left(\rho H_\phi\right) \right)^2 + \frac{1}{\hat{\varepsilon}_r} \left(\frac{\partial H_\phi}{\partial z} \right)^2 \right] \rho d\rho dz
$$
$$
- \pi \int_{z_a}^{z_b} \int_{\rho_a}^{\rho_b} k_0^2 \mu_r H_\phi^2 \rho d\rho dz + 2\pi \int_{z_a}^{z_b} \int_{\rho_a}^{\rho_b} j \frac{k_0}{Z_0} M_\phi H_\phi \rho d\rho dz \quad (2.2)
$$

where z_a, z_b, ρ_a, and ρ_b define the computational domain of the axisymmetric problem in the cylindrical system. Define $J = 2\pi \rho H_\phi$ as current flowing along drill string, the functional based on the new variable is

$$
\Pi(J) = \frac{1}{4\pi} \int_{z_a}^{z_b} \int_{\rho_a}^{\rho_b} \left[\frac{1}{\rho \hat{\varepsilon}_r} \left(\frac{\partial J}{\partial \rho} \right)^2 + \frac{1}{\rho \hat{\varepsilon}_r} \left(\frac{\partial J}{\partial z} \right)^2 \right] d\rho dz
$$
$$
- \frac{1}{4\pi} \int_{z_a}^{z_b} \int_{\rho_a}^{\rho_b} \frac{1}{\rho} k_0^2 \mu_r J^2 d\rho dz + \int_{z_a}^{z_b} \int_{\rho_a}^{\rho_b} j \frac{k_0}{Z_0} M_\phi J d\rho dz \quad (2.3)
$$

The above functional can be discretized and solved by conventional 2D finite elements. However, the efficiency may become rather low if the computational domain becomes very large (e.g. deep well), or the number of formation layers is large. Here we will use a recently developed semianalytical finite element method [5, 9] to solve this axisymmetric electromagnetic problem. Based on the geometric characteristics of the telemetry system, the computational domain will be decomposed into a set of layers, and the structure of each layer is uniform along the vertical direction. 1D finite elements are employed for the discretization of cross section, while the vertical direction is left as undiscretized and will be handled by a Riccati equation based integration scheme [18] with a very high level of accuracy regardless of the thickness of the layer. By transforming this 2D axisymmetric simulation into a set of 1D FEM discretization, the proposed method will lead to much less unknowns and higher efficiency compared with conventional FEM.

Assigning separating planes at the interfaces between layers and at the locations where excitations are imposed, the computational domain will be decomposed into subdomains uniform along the vertical direction. Using 1D finite element for the cross section of a layer and leave the vertical direction untouched, the volume integration part of functional (2.3) will be

$$
\Pi(\mathbf{I}) = \frac{1}{2} \int_{z_a}^{z_b} (\mathbf{J}^T \mathbf{K}_1 \mathbf{J} + \dot{\mathbf{J}}^T \mathbf{K}_2 \dot{\mathbf{J}}) dz \quad (2.4)
$$

where $\dot{\mathbf{J}} = \frac{\partial \mathbf{J}}{\partial z}$, and

$$\mathbf{K}_1 = \sum_e \int_{z_a}^{z_b} \left(\frac{1}{2\pi\rho\hat{\varepsilon}_r} \frac{\partial \mathbf{N}_e}{\partial \rho} \cdot \frac{\partial \mathbf{N}_e^T}{\partial \rho} - \frac{k_0^2 \mu_r}{2\pi\rho} \mathbf{N}_e \cdot \mathbf{N}_e^T \right) dz \qquad (2.5)$$

$$\mathbf{K}_2 = \sum_e \int_{z_a}^{z_b} \left(\frac{1}{2\pi\rho\hat{\varepsilon}_r} \mathbf{N}_e \cdot \mathbf{N}_e^T \right) dz \qquad (2.6)$$

$\mathbf{N}_e$ for the e-th element is a vector containing basis functions as well as testing functions, which can be any interpolatory functions for 1D FEM discretization. Carrying out integration w.r.t. z in the semi-discretized functional (2.4), we will obtain a fully discretized functional as a quadratic function of $\mathbf{J}_a = \mathbf{J}|_{z=z_a}$ and $\mathbf{J}_b = \mathbf{J}|_{z=z_b}$, the values of discretized $\mathbf{J}$ on the upper and lower boundaries of a layer.

$$\Pi(\mathbf{J}_a, \mathbf{J}_b) = \frac{1}{2}\mathbf{J}_a^T \mathbf{K}_{aa} \mathbf{J}_a + \mathbf{J}_b^T \mathbf{K}_{ba} \mathbf{J}_a + \frac{1}{2}\mathbf{J}_b^T \mathbf{K}_{bb} \mathbf{J}_b \qquad (2.7)$$

Matrices $\mathbf{K}_{aa}, \mathbf{K}_{ba}$ can be obtained by applying some numerical integration technique to functional (2.4) w.r.t. the z direction. However, doing this way the accuracy cannot be guaranteed and we lose the chance of exploiting the uniform structural distribution of the layer in the vertical direction. A Riccati equation based high precision integration method [18] can be employed here to deal with the vertical integration with a very high level of accuracy.

It have been proven [19] that system matrices $\mathbf{K}_{aa}, \mathbf{K}_{ba}$, and $\mathbf{K}_{bb}$ can be obtained from

$$\begin{cases} \mathbf{K}_{aa} = -\mathbf{Q} + \mathbf{F}^T \mathbf{G}^{-1} \mathbf{F} \\ \mathbf{K}_{ba} = -\mathbf{G}^{-1} \mathbf{F} \\ \mathbf{K}_{bb} = \mathbf{G}^{-1} \end{cases} \qquad (2.8)$$

And the three matrices $\mathbf{Q}$, $\mathbf{F}$ and $\mathbf{G}$ are solutions to a set of Riccati equations

$$\begin{cases} d\mathbf{F}/d\eta = -\mathbf{G}\mathbf{K}_1 \mathbf{F} = \mathbf{F}\mathbf{K}_2^{-1}\mathbf{Q} \\ d\mathbf{G}/d\eta = \mathbf{K}_2^{-1} - \mathbf{G}\mathbf{K}_1\mathbf{G} = \mathbf{F}\mathbf{K}_2^{-1}\mathbf{F}^T \\ d\mathbf{Q}/d\eta = -\mathbf{F}\mathbf{K}_1\mathbf{F} = \mathbf{Q}\mathbf{K}_2^{-1}\mathbf{Q} - \mathbf{K}_1 \end{cases} \qquad (2.9)$$

with initial conditions

$$\begin{cases} \mathbf{Q}|_{\eta \to 0} = \mathbf{0} \\ \mathbf{G}|_{\eta \to 0} = \mathbf{0} \\ \mathbf{F}|_{\eta \to 0} = \mathbf{I} \end{cases} \qquad (2.10)$$

where $\eta = z_b - z_a$ is the thickness of the layer. $\mathbf{0}$ is a zero matrix, and $\mathbf{I}$ is an identity matrix with the same dimension of $\mathbf{K}_1$ or $\mathbf{K}_2$.

A high precision integration scheme [18] has been developed to solve the above Riccati equations with relative errors as small as machine epsilon on a computer. The first step in this scheme is to divide the integration interval η into 2^N slices

$$\tau = \frac{\eta}{2^N} \tag{2.11}$$

where N is a positive integer number. Under most circumstances, choosing $N = 20$ will make the integration error comparable to machine epsilon defined by double precision. $N = 20$ means $\tau = \eta/1048576$, i.e., even for a layer as thick as 1000 wavelengths, a slice with value τ will be thinner than $1/1000$ of a wavelength. It is sufficiently accurate to use Taylor expansion to calculate matrices $\mathbf{F}$, $\mathbf{G}$, $\mathbf{Q}$ within the small interval τ

$$\begin{cases} \mathbf{F}(\tau) = \mathbf{I} + \mathbf{F}'(\tau) \\ \mathbf{F}'(\tau) = \boldsymbol{\phi}_1\tau + \boldsymbol{\phi}_2\tau^2 + \boldsymbol{\phi}_3\tau^3 + \boldsymbol{\phi}_4\tau^4 + O(\tau^5) \\ \mathbf{G}(\tau) = \boldsymbol{\gamma}_1\tau + \boldsymbol{\gamma}_2\tau^2 + \boldsymbol{\gamma}_3\tau^3 + \boldsymbol{\gamma}_4\tau^4 + O(\tau^5) \\ \mathbf{Q}(\tau) = \boldsymbol{\theta}_1\tau + \boldsymbol{\theta}_2\tau^2 + \boldsymbol{\theta}_3\tau^3 + \boldsymbol{\theta}_4\tau^4 + O(\tau^5) \end{cases} \tag{2.12}$$

where $\boldsymbol{\phi}$, $\boldsymbol{\gamma}$, $\boldsymbol{\theta}$ are matrices with the same dimensions as $\mathbf{F}$, $\mathbf{G}$, $\mathbf{Q}$. Because τ is a very small number, the higher order items $O(\tau^5)$ by Taylor expansion will be smaller than or comparable to machine epsilon on a computer, and dropping off these higher order items will not lead to the loss of any significant digits. A comparison between (2.12) and (2.9) will give expressions of $\boldsymbol{\phi}$, $\boldsymbol{\gamma}$, $\boldsymbol{\theta}$:

$$\begin{cases} \boldsymbol{\gamma}_1 = \mathbf{K}_2^{-1} \\ \boldsymbol{\gamma}_2 = \mathbf{0} \\ \boldsymbol{\gamma}_3 = -\boldsymbol{\gamma}_1\mathbf{K}_1\boldsymbol{\gamma}_1/3 \\ \boldsymbol{\gamma}_4 = \mathbf{0} \end{cases} \tag{2.13}$$

$$\begin{cases} \boldsymbol{\phi}_1 = \mathbf{0} \\ \boldsymbol{\phi}_2 = -\boldsymbol{\gamma}_1\mathbf{K}_1/2 \\ \boldsymbol{\phi}_3 = \mathbf{0} \\ \boldsymbol{\phi}_4 = \left(-\boldsymbol{\gamma}_3\mathbf{K}_1 - \boldsymbol{\gamma}_1\mathbf{K}_1\boldsymbol{\phi}_2\right)/4 \end{cases} \tag{2.14}$$

$$\begin{cases} \boldsymbol{\theta}_1 = -\mathbf{K}_1 \\ \boldsymbol{\theta}_2 = \mathbf{0} \\ \boldsymbol{\theta}_3 = \left(-\boldsymbol{\phi}_2\mathbf{K}_1 - \mathbf{K}_1\boldsymbol{\phi}_2\right)/3 \\ \boldsymbol{\theta}_4 = \mathbf{0} \end{cases} \tag{2.15}$$

And the solutions of three matrices over interval 2τ, i.e. $\mathbf{F}(2\tau)$, $\mathbf{G}(2\tau)$, $\mathbf{Q}(2\tau)$ will be

$$\begin{cases} \mathbf{G}(2\tau) = \mathbf{G}(\tau) + \mathbf{F}(\tau)[\mathbf{G}(\tau)^{-1} + \mathbf{Q}(\tau)]^{-1}\mathbf{F}(\tau)^T \\ \mathbf{F}'(2\tau) = \mathbf{F}'(\tau)[\mathbf{I} + \mathbf{G}(\tau)\mathbf{Q}(\tau)]^{-1}\mathbf{F}'(\tau) \\ \qquad + [(\mathbf{F}'(\tau) - \mathbf{G}(\tau)\mathbf{Q}(\tau)/2][\mathbf{I} + \mathbf{G}(\tau)\mathbf{Q}(\tau)]^{-1} \\ \qquad + [\mathbf{I} + \mathbf{G}(\tau)\mathbf{Q}(\tau)]^{-1}[\mathbf{F}'(\tau) - \mathbf{G}(\tau)\mathbf{Q}(\tau)/2] \\ \mathbf{Q}(2\tau) = \mathbf{Q}(\tau) + \mathbf{F}(\tau)^T[\mathbf{Q}(\tau)^{-1} + \mathbf{G}(\tau)]^{-1}\mathbf{F}(\tau) \end{cases} \quad (2.16)$$

It should be noted that only the increment $\mathbf{F}'(\tau) = \mathbf{F}(\tau) - \mathbf{I}$ instead of $\mathbf{F}(\tau)$ itself is calculated and stored during the above step. The reason that the small quantity must be kept away from $\mathbf{I}$ during computation, otherwise significant digits of $\mathbf{F}'(\tau)$ will be lost if add a small quantity $\mathbf{F}'(\tau)$ to a much larger quantity $\mathbf{I}$. Repeating (2.16) for N times, the integration interval will grow to η, and we then can obtain the matrices $\mathbf{G}(\eta)$, $\mathbf{Q}(\eta)$, and $\mathbf{F}(\tau) = \mathbf{F}'(\tau) + \mathbf{I}$ with a very high level of accuracy. Table 2.1 lists relative errors of the high precision integration for a simple case as a plane wave traveling through a distance of one wavelength in a homogeneous medium. We can see that when N is equal to or larger than 14, the relative error is close to 2.22×10^{-16}, which is the machine epsilon defined by double precision. This table suggests that to achieve a higher accuracy or efficiency, N in the proposed algorithm can be chosen adaptively based on the thickness of each layer along the z direction, instead of being fixed as $N = 20$.

After performing the high precision integration in the vertical direction, the final system matrix of this layer will be as

$$\mathbf{K}_e = \begin{bmatrix} \mathbf{K}_{aa} & \mathbf{K}_{ba}^T \\ \mathbf{K}_{ba} & \mathbf{K}_{bb} \end{bmatrix} \quad (2.17)$$

At this stage this discretized layers is ready to be assembled with adjacent layers, discretized by either conventional finite elements or by the proposed semianalytical elements. After the global system matrix is formed, and the sources terms is assigned to the right hand sides, numerical results will be obtained by solving the system of

Table 2.1 Relative error of high precision integration w.r.t. different value of N

Value of N	Integration points (2^N)	Relative error
2	4	2.7776e−01
4	16	1.2431e−03
6	64	4.8636e−06
8	256	1.9000e−08
10	1024	7.4220e−11
12	4096	2.9019e−13
14	16384	6.4325e−16
16	65536	2.4493e−16
18	262144	2.0213e−15
20	1048576	2.4493e−16

equations and based on which the performance of the electromagnetic telemetry system can be evaluated. Oftentimes the formation for EM telemetry study can be regarded as a layered earth model based on resistivity log. The system matrix by the semianalytical FEM for a layered structure will be in a block tridiagonal form [4], which can be efficiently solved by the block Thomas algorithm [7, 14] with the computational cost linearly proportional to the number of layers. The combination of semianalytical FEM with the block Thomas algorithm is promising in modeling EM telemetry in unground formation with a large number of layers.

2.3 Examples and Discussions

We first consider a well in a homogeneous underground formation with resistivity $1\,\Omega\,m$. The total length of the drill string is 5000 ft. The diameters of the drill pipe and the borehole are 10 inch and 12 inch, respectively. The length of casing string as well as the cement is 3000 ft. The resistivity of drilling fluid filled in borehole is set as $1\,\Omega\,m$. The conductivities of both drilling pipe and casing string are assumed as 2e6 S/m. The downhole transmitter as a constant 1 V voltage source is 200 ft behind the drill bit transmitting signals at 5 Hz working frequency, and the surface receiver is measuring the voltage difference between the BOP and a point 500 ft away from the rig. Figure 2.2 shows the calculated amplitude of current flowing along the drill string. We can see that the amplitude of the current gradually decreases as it is away from the source, and this is because the current keeps leaking into conductive formation while flowing along the drill string. From this figure we can also observe that the current magnitude deceases faster since the depth of 3000 ft and upwardly,

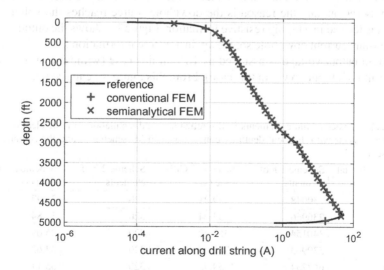

Fig. 2.2 Amplitude of current flow along the drill string in a homogeneous formation model

and this is due to the existence of steel casing. The current flow along the drill pipe calculated by both conventional FEM and the proposed semianalytical FEM are shown in Fig. 2.2. This figure also shows the reference results, which is obtained by conventional FEM with a very fine mesh. We can see that the results by both numerical schemes agree well with the reference. A quantitative comparison between the two schemes suggests that to achieve a similar level of accuracy (in this case relative errors are 9.71e−3 and 7.46e−3 for conventional FEM and semianalytical FEM, respectively), the semianalytical FEM costs much less computational time (34.57 s) than does the conventional FEM (302.10 s). A laptop with an Intel Core i7-4600U CPU at 2.1 GHz and 16 GB memory was used in this case and all the following examples in this chapter.

Table 2.2 lists the computational costs of electromagnetic telemetry systems for wells with different depth by the two numerical schemes. We can see that the computational cost by the conventional FEM increases rapidly as the well goes deeper. This is because a deeper well means a larger domain of simulation, and consequently more elements in discretization as well as a larger system of equations to solve. On the other hand, the computational time by the semianalytical FEM stays about the same regardless of the depth of the well. The reason is that semianalytical FEM only requires discretization on the cross section of each layer, and this scheme can handle the vertical integration very accurately no matter how thin or how thick the layer is.

The three most significant factors deciding the EM telemetry signal strength are working frequency, resistivity or conductivity of underground formation, and drilling depth. Figure 2.3 shows simulated results of the signal strength of an electromagnetic telemetry signal w.r.t. different system parameters. We can see that the strength of telemetry signal decreases rapidly as the drilling goes deeper or formation becomes more conductive, or if a higher working frequency is employed for faster data transmission. In this figure we also notice that all the curves become saturated in the high resistivity region, and the reason is the downhole source reaches the voltage limit when it is located in a highly resistive formation. Figure 2.4 shows the output voltage of a constant current downhole source against different formation resistivity, and the maximum output voltage is assumed as 18 V in this case (without considering the upper limit of output power in this case). From this figure we can see that the higher

Table 2.2 Number of finite elements in discretization, and simulation time of electromagnetic telemetry systems with different depth by conventional FEM (scheme 1) and the semianalytical FEM (scheme 2)

Depth of drill bit (ft)	Scheme 1 # of elements	Scheme 1 CPU time (s)	Scheme 2 # of elements	Scheme 2 CPU time (s)
2000	130048	59.21	320	33.97
4000	236966	233.64	320	33.84
6000	340154	353.32	320	35.09
8000	539963	787.50	320	34.63
10000	667234	1259.93	320	35.11

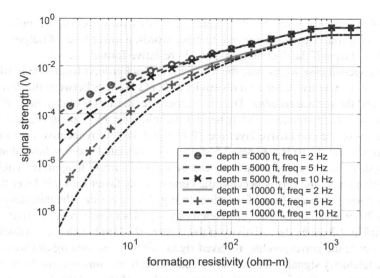

Fig. 2.3 Telemetry signals with different working frequency, drilling depth, and formation resistivity

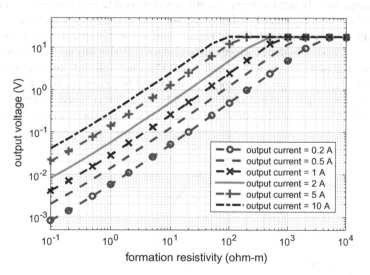

Fig. 2.4 The output voltage of a constant current downhole source w.r.t. underground formation resistivity. The maximum output voltage in this case is set as 18 V

the level of constant current the downhole source uses, the lower the formation resistivity for the source to turn saturated. In other words, a lower level of output current should be chosen if the drill bit is in a highly resistive formation.

In the second case we assume the resistivity of background formation is $10\,\Omega\,m$, and a more conductive layer with resistivity as $0.1\,\Omega\,m$ exists between the downhole source and the surface receiver. The true vertical depth (TVD) of drill bit is 8000 ft, and the downhole source is 200 ft behind the bit. The TVD of upper and lower boundaries of the conductive layer are 3900 ft and 4100 ft, respectively. Operating frequency is set as 10 Hz. Figure 2.5 shows the simulated current flow along the drill pipe. We can see that there is a steep descent of current magnitude within the conductive layer, and this is because the current leaks faster into this layer than in the more resistive background formation. In Fig. 2.6 we assume the thickness of the layer is 200 ft, and plot the strength of telemetry signal w.r.t. different resistivity of the layer. In Fig. 2.7 we fix the resistivity of the layer as $0.1\,\Omega\,m$ but change its thickness, and calculate the corresponding received signal. From these two figures we can see that the telemetry signal will be greatly weakened if it encounters a conductive layer during the propagation from downhole source to the surface, and this is a limitation of the electromagnetic telemetry compared with mud pulse technique. In an improved electromagnetic telemetry system recently developed, one terminal of the receiver is be extended to the bottom of casing string and connected to the casing shoe instead of a point on the surface [6]. Heavy signal attenuation due to conductive formation between casing shoe and the surface can be circumvented in this novel design, and the picked up telemetry signal will be enhanced. Detailed descriptions and discussions of this novel electromagnetic telemetry system will be given in Chap. 4.

The third example is a field job done in Oklahoma. Water based drilling fluid with $1\,\Omega\,m$ resistivity was used in this case. The working frequency and output current

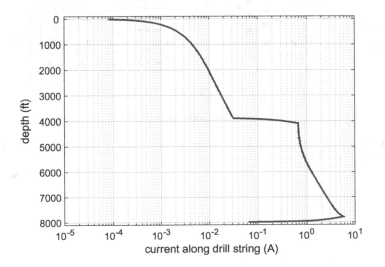

Fig. 2.5 Amplitude of current flow along the drill string in a layered formation model

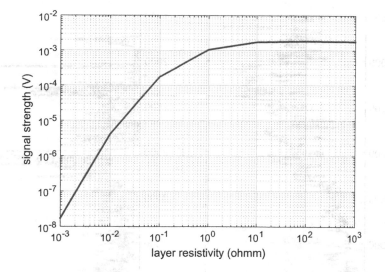

Fig. 2.6 Telemetry signals w.r.t. resistivity of a 200 ft thick layer between bit and surface

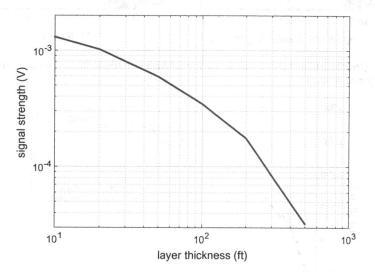

Fig. 2.7 Telemetry signals w.r.t. thickness of a 0.1 Ω m layer between bit and surface

of the downhole source were set as 5.33 Hz and 1 A, respectively. The diameters of drill pipe and of the borehole are 6.75 inches and 11 inches, respectively. Figure 2.8 shows the resistivity log and noise record of this job from 1500 ft to 4500 ft. The measured telemetry signal and simulated results are shown in Fig. 2.9, from which we can see that the calculated strength of EM telemetry signal follows well with the trends of the field measurements in a large range of depth. It should be noted that a lot of uncertainties in real cases will lead to discrepancies between numerical

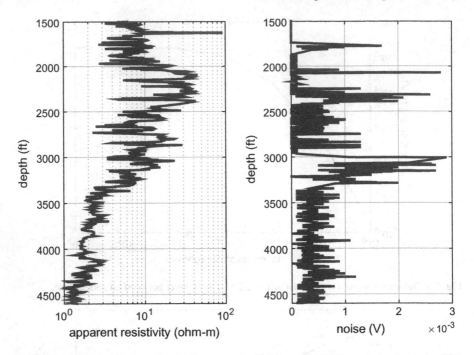

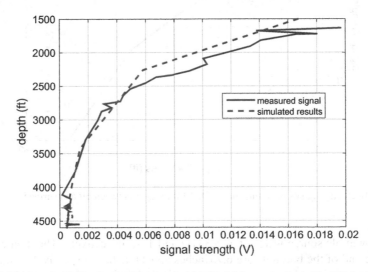

Fig. 2.8 (left) Resistivity log of a wireline survey in Oklahoma; (right) Recorded noise during EM telemetry measurement

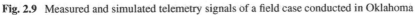

Fig. 2.9 Measured and simulated telemetry signals of a field case conducted in Oklahoma

simulation results and actual measurements. For example, the resistivity distribution near the surface is subject to great variation due to temperature and moisture; the mud resistivity in borehole may change with depth; and the formation resistivity may have lateral variation in each layer, leading the whole system to be a 3D structure instead of an axisymmetric problem. After all, EM telemetry is used in subsurface wireless communication, not as a quantitative well logging tool. A very high level of accuracy may not be the first priority for modeling of EM telemetry, but an efficiently simulation method as proposed in this paper will greatly help the evaluation and decision-making of EM telemetry technique for specific field jobs.

References

1. Bhagwan, J., Trofimenkoff, F.: Electric drill stem telemetry. IEEE Trans. Geosci. Remote Sens. **2**, 193–197 (1982)
2. Carcione, J.M., Poletto, F.: A telegrapher equation for electric-telemetering in drill strings. IEEE Trans. Geosci. Remote Sens. **40**(5), 1047–1053 (2002)
3. Carcione, J.M., Poletto, F.: Electric drill-string telemetry. J. Comput. Phys. **186**(2), 596–609 (2003)
4. Chen, J.: An efficient discontinuous galerkin finite element method with nested domain decomposition for simulations of microresistivity imaging. J. Appl. Geophys. **114**, 116–122 (2015)
5. Chen, J.: A semianalytical finite element analysis of electromagnetic propagation in stratified media. Microw. Opt. Technol. Lett. **57**(1), 15–18 (2015)
6. Chen, J., Li, S., MacMillan, C., Cortes, G., Wood, D., et al.: Long range electromagnetic telemetry using an innovative casing antenna system. In: SPE Annual Technical Conference and Exhibition. Society of Petroleum Engineers (2015)
7. Chen, J., Tobon, L.E., Chai, M., Mix, J.A., Liu, Q.H.: Efficient implicit-explicit time stepping scheme with domain decomposition for multiscale modeling of layered structures. IEEE Trans. Compon. Packag. Manuf. Technol. **1**(9), 1438–1446 (2011)
8. Chen, J., Zeng, S., Dong, Q., Huang, Y.: Rapid simulation of electromagnetic telemetry using an axisymmetric semianalytical finite element method. J. Appl. Geophys. **137**, 49–54 (2017)
9. Chen, J., Zhu, B., Zhong, W., Liu, Q.H.: A semianalytical spectral element method for the analysis of 3-d layered structures. IEEE Trans. Microw. Theory Tech. **59**(1), 1–8 (2011)
10. DeGauque, P., Grudzinski, R., et al.: Propagation of electromagnetic waves along a drillstring of finite conductivity. SPE Drill. Eng. **2**(02), 127–134 (1987)
11. Hill, D., Wait, J.: Electromagnetic basis of drill-rod telemetry. Electron. Lett. **14**(17), 532–533 (1978)
12. Li, W., Nie, Z., Sun, X.: Wireless transmission of mwd and lwd signal based on guidance of metal pipes and relay of transceivers. IEEE Trans. Geosci. Remote Sens. **54**(8), 4855–4866 (2016)
13. Li, W., Nie, Z., Sun, X., Chen, Y.: Numerical modeling for excitation and coupling transmission of near field around the metal drilling pipe in lossy formation. IEEE Trans. Geosci. Remote Sens. **52**(7), 3862–3871 (2014)
14. Meurant, G.: A review on the inverse of symmetric tridiagonal and block tridiagonal matrices. SIAM J. Matrix Anal. Appl. **13**(3), 707–728 (1992)
15. Soulier, L., Lemaitre, M., et al.: EM MWD data transmission status and perspectives. In: SPE/IADC Drilling Conference. Society of Petroleum Engineers (1993)
16. Vong, P., Lai, H., Rodger, D.: Modeling electromagnetic field propagation in eddy-current regions of low conductivity. IEEE Trans. Magn. **42**(4), 1267–1270 (2006)

17. Vong, P.K., Rodger, D., Marshall, A.: Modeling an electromagnetic telemetry system for signal transmission in oil fields. IEEE Trans. Magn. **41**(5), 2008–2011 (2005)
18. Zhong, W.X.: On precise integration method. J. Comput. Appl. Math. **163**(1), 59–78 (2004)
19. Zhong, W.X.: Duality system in applied mechanics and optimal control, vol. 5. Springer Science & Business Media (2006)

Chapter 3
3D Modeling of Electromagnetic Telemetry

3.1 Electromagnetic Telemetry in Directional and Horizontal Drilling

Unlike the EMT in vertical drilling discussed in Chap. 2, which can be simplified as an axisymmetric structure, EMT for directional and horizontal drilling are complicate 3D systems. More specifically, as one compares the spatial dimensions of the entire computational domain (tens of thousands of feet) and the fine details such as the gap source in the borehole (a fraction of an inch), the EMT system can be considered as an extremely multiscale structure. Besides, the earth model in the computational domain consists of a large number of formation layers with different electromagnetic properties and oftentimes anisotropy. Using conventional numerical methods such as FEM or FDM to discretize such a complicate and multiscale structure will easily lead to millions of or a even bigger number of unknowns. Recently, a special FEM has been proposed to augment the usual volume-based conductivity on tetrahedra by facet- and edge-based conductivity on the infinitesimally thin regions between elements, thus avoiding the costly discretization of the thin structures [27, 31]. However, this method is only feasible at static frequency.

The integral equation method (IEM) avoids the discretiztion of the background formation, resulting a much smaller number of unknowns [3, 7, 10, 24, 26, 30, 32, 33]. Different geometries and formations have been studied, e.g., modeling of the EMT systems for vertical and horizontal drilling in homogeneous [3, 10], half-space [7, 24, 30], and layered media [26, 32, 33]. The IEM scheme only needs to discretize the surface or the volume of the scatterer, e.g., the drill string and the borehole, resulting a much easier discretization and faster solution. Further reduction of unknowns can be attained by the thin wire model that is qualified for wire structure having radius much smaller than its length. On the other hand, the main drawback of IEM is that this method generates a full dense system matrix and will encounter the unavoidable singularities residing in the Green's function and sometimes the basis function. The full dense system matrix is not a big issue for the modeling of the

© The Author(s), under exclusive license to Springer Nature Switzerland AG 2019
J. Chen et al., *Borehole Electromagnetic Telemetry System*,
SpringerBriefs in Petroleum Geoscience & Engineering,
https://doi.org/10.1007/978-3-030-21537-8_3

EMT system since the total number of the unknowns is not large when the thin wire approximation is applied. Also, the singularity issue can be resolved by singularity substraction scheme [9] and singularity cancellation scheme [28]. Although in theory the IEM scheme is suitable for the simulation of the EMT system for general 3D well trajectories, extension of the IEM scheme to the directional and horizontal drilling in layered TI media have not been discussed and implemented. One major challenge is the efficient and robust evaluation of the Green's function for layered TI formation. In this chapter we will elaborate the IEM scheme for thin wire model in layered TI media, and its applications to EMT modeling in directional and horizontal drilling.

3.2   IEM for Thin Wire Model in Layer TI Media

As shown in Fig. 3.1, an EMT system has been deployed for directional drilling inside a horizontally layered transversely isotropic (TI) formation. Each formation layer has horizontal and vertical conductivities $(\sigma_{hi}, \sigma_{vi})$ $(i = 1, 2, \ldots, N)$. IEM with thin wire model is utilized to efficiently model EMT in directional and horizontal drilling, which is always a 3D multiscale electromagnetic problem.

Fig. 3.1 A schematic of the EMT system for directional and horizontal drilling in a layered TI formation

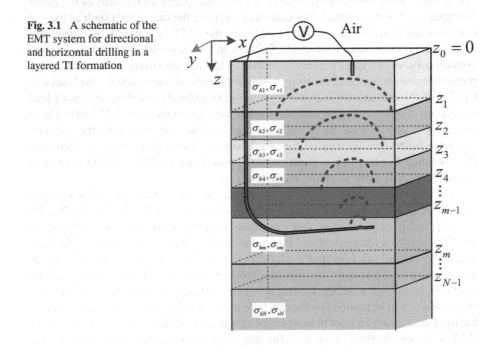

3.2.1 Thin Wire in Homogeneous Isotropic Media

To model the electromagnetic fields excited by the voltage or current gap source, we apply the EFIE in the frequency domain

$$\left[\mathbf{E}^i(\mathbf{r}) + \mathbf{E}^s(\mathbf{r})\right]_{tan} = Z_i \mathbf{I}(\mathbf{r}), \quad \mathbf{r} \in S \tag{3.1}$$

to represent the total tangential electric field on the surface of the drill string that is represented by S. $\mathbf{I}(\mathbf{r})$ is the total current along the drill string which is assumed to be axially directed when the drill string is very thin. Z_i denotes the internal impedance of the drill string and characterizes the effect of the finite conductivity of the drill string. If we assume the drill string is a perfect conductor, then Z_i vanishes. If the conductivity of the drill string is finite, e.g., σ_d, we can assign a specific value to Z_i as [7]

$$Z_i = \frac{\gamma}{2\pi a \sigma_d} \frac{J_0(\gamma r_e)}{J_1(\gamma r_e)} \tag{3.2}$$

where $\gamma = \sqrt{\omega \mu_0 (\varepsilon_0 - j\sigma_s/\omega)}$ and a is the radius of the drill string. J_0 and J_1 are the zero- and first-order Bessel functions, respectively. The incident electric field $\mathbf{E}^i(\mathbf{r})$ is a voltage/current gap source, as shown in Fig. 3.1. Using the mixed-potential form of electric field, the scattering electric field $\mathbf{E}^s(\mathbf{r})$ can be expressed in terms of the magnetic vector potential $\mathbf{A}(\mathbf{r})$ and scalar potential $\Phi(\mathbf{r})$ as

$$\mathbf{E}^s(\mathbf{r}) = -j\omega \mathbf{A}(\mathbf{r}) - \nabla \Phi(\mathbf{r}), \tag{3.3}$$

where $\omega = 2\pi f$ and f is the frequency of the voltage/current gap source.

Firstly, consider the case where the background formation is homogeneous isotropic media with the wavenumber

$$k = \omega \sqrt{\varepsilon_0 \varepsilon_{rc} \cdot \mu_0 \mu_r} \tag{3.4}$$

and

$$\varepsilon_{rc} = \varepsilon_r - j\frac{\sigma}{\omega \varepsilon_0}, \tag{3.5}$$

where ε_0, ε_r, ε_{rc} σ, μ_0, and μ_r denote the permittivity of free space, the relative permittivity of the media, the relative effective permittivity of the media, the conductivity of the media, the permeability of free space, and the relative permeability of the media, respectively.

The EMT system needs to operate at very low frequency from a fraction of a hertz to tens of hertz, otherwise the surface signal will be not detectable due to the heavy attenuation of electromagnetic fields passing through the formation. Hence, the drill string has an very electrically small radius compared to the wavelength of the formation. So that it can be treated using the thin wire approximation that assumes the current density on the thin wire has no azimuthal component. The axially directed current on the thin wire also has no variation along the azimuthal angle. There is no radial component since the current only flows on the surface of the drill string. Accordingly, the induced surface current on the wire (drill string) is assumed to be axially directed and has no azimuthal variation [20]. Furthermore, the potentials can be represented as functions of the total induced axial current $\mathbf{I}(\mathbf{r})$ on S in a simple way, i.e.,

$$\mathbf{A}(\mathbf{r}) = \mu_0 \mu_r \int_S G(\mathbf{r}, \mathbf{r}') \cdot \frac{\mathbf{I}(\mathbf{r}')}{2\pi a(\mathbf{r}')} dS' \tag{3.6}$$

and

$$\Phi(\mathbf{r}) = -\frac{1}{j\omega\varepsilon_0\varepsilon_{rc}} \int_S G(\mathbf{r}, \mathbf{r}') \nabla'_s \cdot \left(\frac{\mathbf{I}(\mathbf{r}')}{2\pi a(\mathbf{r}')} \right) dS', \tag{3.7}$$

where $a(\mathbf{r}')$ is the radius of the drill string cross section which includes the point $\mathbf{r}'$ on the wire axis. Here, the prime symbols denote the position of the source. In the above integral equations,

$$G(\mathbf{r}, \mathbf{r}') = \frac{e^{-jkR}}{4\pi R}, \quad R = |\mathbf{r} - (\mathbf{r}' + \mathbf{a}(\mathbf{r}'))| \tag{3.8}$$

is the Green's function for homogeneous isotropic media. The vector $(\mathbf{r}' + \mathbf{a}(\mathbf{r}'))$ is a point on the surface S, where $\mathbf{a}(\mathbf{r}')$ is a vector with magnitude of $a(\mathbf{r}')$ and points from $\mathbf{r}'$ to the surface of the drill string.

Since the current on the surface of drill string has no azimuthal variation, the so-called thin wire kernel [28]

$$K(\mathbf{r}, \mathbf{r}') = \frac{1}{2\pi} \int_{-\pi}^{\pi} G(\mathbf{r}, \mathbf{r}') d\phi', \tag{3.9}$$

can be extracted from the potential integrals (3.6) and (3.7). This thin wire kernel helps reduce the surface integrals into line integrals with the form

$$\int_L B(\mathbf{r}') K(\mathbf{r}, \mathbf{r}') dl' \tag{3.10}$$

where $B(\mathbf{r}')$ is the total scalar current or charge on the circumference of the corresponding cross section and L is the axial length of the thin wire. Accurate evaluation of the thin wire kernel is vital to model the thin wire structure, especially when observation point $\mathbf{r}$ is near to and coincides with the source point $\mathbf{r}'$. Fortunately, the potential integrals on the thin wire can be calculated with high accuracy when singularity cancellation schemes are employed [28].

3.2.2 Green's Function for Layered TI Media

Now consider a multi-layered media with TI media for each layer, as shown in Fig. 3.1. For each layer, the TI conductivity tensor $\sigma = \mathcal{I}_t \sigma_t + \hat{z}\hat{z}\sigma_z$ is incorporated into the relative effective or complex permittivity $\varepsilon_{cr} = \mathcal{I}_t \varepsilon_{crt} + \hat{z}\hat{z}\varepsilon_{crz}$ by the formulas $\varepsilon_{crt} = \varepsilon_{rt} - j\frac{\sigma_t}{\omega \varepsilon_0}$ and $\varepsilon_{crz} = \varepsilon_{rz} - j\frac{\sigma_z}{\omega \varepsilon_0}$, where $\mathcal{I}_t$ is the transverse unit dyadic, and the relative permeability of TI media is $\mu_r = \mathcal{I}_t \mu_{rt} + \hat{z}\hat{z}\mu_{rz}$.

The well-known time-harmonic Maxwell's equations are

$$\nabla \times \mathbf{E} = -j\omega\mu_0\mu_r \cdot \mathbf{H} - \mathbf{M}, \tag{3.11}$$

and

$$\nabla \times \mathbf{H} = j\omega\varepsilon_0\varepsilon_{cr} \cdot \mathbf{E} + \mathbf{J}, \tag{3.12}$$

where $\mathbf{J}$ and $\mathbf{M}$ are the electric and magnetic current sources.

Because the Green's function is the solution to a dipole source, it is never a trivial task to acquire the close-form Green's function for layered TI media in the spatial domain. However, resorting to the symmetric formation w.r.t. the vertical axis, we can express any vector component as $\mathbf{F}(\mathbf{r}) \equiv \mathbf{F}(\rho, z)$, where $\rho = \hat{x}x + \hat{y}y$ is the projection of $\mathbf{r}$ on the $x - y$ plane. Then, we define a 2D Fourier transformation for all the fields w.r.t. to the transverse plane to facilitate the analysis of layered TI media. After the Fourier transformation, the layered media problem goes into the spectral domain or the wavenumber domain. The physical meaning of Fourier transformation is to convert a dipole source in the spatial domain into an infinity series of plane waves in the spectral domain. After Fourier transformation, the modeling of plane waves propagating in layered TI media can be solved using the transmission line analogy [17].

The analogy of transmission line facilitates to obtain the mixed-potential form of LMGF in Dyadic form [17]. For thin wire structure in layered TI media, the vector and scalar potentials relate to the electric current source and LMGF via

$$\mathbf{A}(\mathbf{r}) = \mu_0 \int_S \mathcal{G}^A(\mathbf{r}, \mathbf{r}') \cdot \frac{\mathbf{I}(\mathbf{r}')}{2\pi a(\mathbf{r}')} dS', \tag{3.13}$$

$$\Phi(\mathbf{r}) = -\frac{1}{j\omega\varepsilon_0}\left[\int_S K^\Phi(\mathbf{r}, \mathbf{r}')\nabla'_s \cdot \frac{\mathbf{I}(\mathbf{r}')}{2\pi a(\mathbf{r}')} dS' + \int_S P_z(\mathbf{r}, \mathbf{r}')\hat{z} \cdot \frac{\mathbf{I}(\mathbf{r}')}{2\pi a(\mathbf{r}')} dS'\right]. \tag{3.14}$$

Components of the dyadic Green's function $\mathcal{G}^A$ physically represent the components of the magnetic vector potential at $\mathbf{r}$ for the variously-oriented unit-strength electric current dipoles at $\mathbf{r}'$. K^Φ is the corresponding scalar potential kernel, and P_z is the vertical current scalar potential correction factor for layered media. All the Green's functions are called the mixed-potential form of LMGF or MP-LMGF. Similar formulas can be acquired for magnetic dipole source using the duality theory. For conciseness, they are not discussed here and in the following.

These dyadic and scalar LMGF's do not have a close form in spatial domain and need to be calculated from their corresponding spectral form with the help of inverse Fourier transformation

$$
\begin{bmatrix} \mathcal{G}^A \\ K^\Phi \\ P_z \end{bmatrix} = \frac{1}{(2\pi)^2} \int\limits_{-\infty}^{\infty} \int\limits_{-\infty}^{\infty} \begin{bmatrix} \tilde{\mathcal{G}}^A(k_\rho, z, z') \\ \tilde{K}^\Phi(k_\rho, z, z') \\ \tilde{P}_z(k_\rho, z, z') \end{bmatrix} e^{-j\mathbf{k}_\rho \cdot (\rho - \rho')} dk_x dk_y, \tag{3.15}
$$

where the functions in the square brackets at the right side are the corresponding spectral components of LMGF's that can be expressed using the analogy of transmission line voltages and currents. More distinctly, these LMGF's in the spectral domain are

$$
\tilde{\mathcal{G}}^A(\mathbf{k}_\rho, z, z') = \begin{bmatrix} \frac{1}{j\omega\mu_0} V_i^h & 0 & 0 \\ 0 & \frac{1}{j\omega\mu_0} V_i^h & 0 \\ \frac{\mu_t k_x}{j k_\rho^2}(I_i^h - I_i^e) & \frac{\mu_t k_y}{j k_\rho^2}(I_i^h - I_i^e) & \frac{\mu_t}{j\omega\varepsilon_0\varepsilon'_{cz}} I_v^e \end{bmatrix}, \tag{3.16}
$$

$$
\tilde{K}^\Phi(\mathbf{k}_\rho, z, z') = j\omega\varepsilon_0 \frac{V_i^e - V_i^h}{k_\rho^2}, \tag{3.17}
$$

and

$$
\tilde{P}_z(\mathbf{k}_\rho, z, z') = \frac{k_0^2 \mu'_r}{k_\rho^2}(V_v^h - V_v^e). \tag{3.18}
$$

where V and I are the voltage and current on the transmission line model of the layered structure. The definitions and expressions for the voltages and currents appearing in the spectral form can be found in [17]. The unprimed material parameters in (3.16)–(3.18) are evaluated at the observation point $\mathbf{r}$, while the primed parameters are evaluated at the source point $\mathbf{r}'$.

After the LMGF's are obtained in the spectral domain, we are ready to calculate the LMGF's in spatial domain using (3.15). To save the computation effort, the Hankel transform

$$
\frac{1}{(2\pi)^2} \int\limits_{-\infty}^{\infty} \int\limits_{-\infty}^{\infty} \begin{bmatrix} \sin n\varphi \\ \cos n\varphi \end{bmatrix} \tilde{F}(k_\rho) e^{-j\mathbf{k}_\rho \cdot (\rho - \rho')} dk_x dk_y = (-j)^n \begin{bmatrix} \sin n\gamma \\ \cos n\gamma \end{bmatrix} S_n \left\{ \tilde{F}(k_\rho) \right\}
$$

$$
\tag{3.19}
$$

where $\varphi = \tan^{-1}(k_y/k_x)$ and $\gamma = \tan^{-1}(y - y')/(x - x')$, is applied to reduce the double infinite integral into a single semi-infinite integral. With a rigorous derivation, the calculation of all the components of LMGF's involving general electric and magnetic sources can be expressed in terms of 14 independent integrals [14]

$$I_1 = S_0 \{V_i^h\},$$

$$I_2 = S_0 \{I_v^e\},$$

$$I_3 = S_0 \{I_i^h + I_i^e\},$$

$$I_4 = S_0 \{V_v^e + V_v^h\},$$

$$I_5 = S_0 \left\{ \frac{V_i^h - V_i^e}{k_\rho^2} \right\},$$

$$I_6 = S_0 \left\{ \frac{V_v^h - V_v^e}{k_\rho^2} \right\},$$

$$I_7 = S_0 \left\{ \frac{I_v^e - I_v^h}{k_\rho^2} \right\},$$

$$I_8 = S_0 \left\{ \frac{I_i^e - I_i^h}{k_\rho^2} \right\},$$

$$I_9 = S_1 \left\{ \frac{I_i^e - I_i^h}{k_\rho} \right\},$$

$$I_{10} = S_1 \left\{ \frac{V_v^h - V_v^e}{k_\rho} \right\},$$

$$I_{11} = S_1 \{k_\rho V_i^h\},$$

$$I_{12} = S_1 \{k_\rho I_v^e\},$$

$$I_{13} = S_2 \{I_i^e - I_i^h\},$$

$$I_{14} = S_2 \{V_v^h - V_v^e\}. \tag{3.20}$$

All of them are in the form of the generalized Sommerfeld integral (SI)

$$S_n\{\tilde{F}(k_\rho)\} = \frac{1}{2\pi} \int_0^\infty \tilde{F}(k_\rho) J_n(k_\rho |\boldsymbol{\rho} - \boldsymbol{\rho}'|) k_\rho dk_\rho. \tag{3.21}$$

Here, J_n is the Bessel function of the first kind of order n with $n = 0, 1, 2$ and ρ, ρ' are the cylindrical coordinates of the projections of the observation and source points on the transverse plane.

From (3.16)–(3.18), we recognize that one needs to calculate integrals I_1, I_2, I_5, and I_6 if presenting the electric current source. The algorithms for the accurate and efficient evaluation of these independent SI's involve the deformed integral path [1], acceleration and regularization of SIs by asymptotic singularity extraction [5, 23], and weighted average method (WAM) for the integral tails [16, 18].

The generalized SI's in (3.20) do not converge rapidly for small vertical displacements between source and observation points. The convergence rate of these integrals can be significantly accelerated by extracting asymptotic terms from the integrand. These terms are generally interpreted as direct radiation and reflection from lower and upper boundaries of the source layer when source and observation points are in the same layer, as shown in Fig. 3.2. When source and observation points are in two adjacent layers, the extracted term is only the direct term. When source and observation points are separated by more than one layer, the exponential decay of the quantities along the transmission line guarantees sufficiently fast convergence of the integrals, and no extraction is required. The extraction of direct and *quasi-static* image (reflection) terms from the integrands to accelerate the integrals simultaneously regularizes the spectral contributions [13, 14, 23]. The remaining integrals behave much smoother when the observation point is near the source point, achieving a much faster convergence for the evaluation of the SI's.

The asymptotic singularity extraction can be generally expressed as

$$S_n \left\{ \tilde{F}(k_\rho) \right\} = S_n \left\{ \tilde{F}(k_\rho) - \tilde{F}^\infty(k_\rho) \right\} + F^\infty(\mathbf{r}, \mathbf{r}') \tag{3.22}$$

with $\tilde{F}^\infty(k_\rho)$ being the asymptotic form of $\tilde{F}(k_\rho)$. The extracted term can be expressed by their closed-form $F^\infty(\mathbf{r}, \mathbf{r}')$ in spatial domain with the use of the following Sommerfeld and related identities [2, 8]

$$S_0 \left\{ \frac{e^{-jk_z^\alpha |\zeta|}}{2jk_z^\alpha} \right\} = \frac{1}{\nu^\alpha} \frac{e^{-jk_t R^\alpha}}{4\pi R^\alpha},$$

$$S_0 \left\{ \frac{e^{-jk_z^\alpha |\zeta|}}{2} \right\} = \frac{1}{\nu^\alpha} |\zeta| (1 + jk_t R^\alpha) \frac{e^{-jk_t R^\alpha}}{4\pi (R^\alpha)^3},$$

$$S_0 \left\{ \frac{e^{-jk_z^\alpha |\zeta|}}{2(jk_z^\alpha)^2} \right\} = \frac{1}{\nu^\alpha} \int\limits_{|\zeta|}^\infty \frac{e^{-jk_t \sqrt{\rho^2/\nu^\alpha + z^2}}}{4\pi \sqrt{\rho^2/\nu^\alpha + z^2}} dz \equiv \frac{1}{\nu^\alpha} G_z^\alpha(\rho, |\zeta|),$$

$$S_0 \left\{ \frac{e^{-jk_z^\alpha |\zeta|}}{2(jk_z^\alpha)^3} \right\} = -\frac{1}{\nu^\alpha} \left[j \frac{e^{-jk_t R^\alpha}}{4\pi k_t} + |\zeta| G_z^\alpha(\rho, |\zeta|) \right],$$

$$S_1 \left\{ \frac{e^{-jk_z^\alpha |\zeta|}}{2k_\rho} \right\} = \frac{R^\alpha e^{-jk_t |\zeta|} - |\zeta| e^{-jk_t R^\alpha}}{4\pi \rho R^\alpha},$$

$$S_1 \left\{ \frac{k_\rho e^{-jk_z^\alpha |\zeta|}}{2jk_z^\alpha} \right\} = \frac{1}{(\nu^\alpha)^2} \rho [1 + jk_t R^\alpha] \frac{e^{-jk_t R^\alpha}}{4\pi (R^\alpha)^3},$$

$$S_2 \left\{ \frac{e^{-jk_z^\alpha |\zeta|}}{2} \right\} = \frac{2}{\rho} S_1 \left\{ \frac{e^{-jk_z^\alpha |\zeta|}}{2k_\rho} \right\} - S_0 \left\{ \frac{e^{-jk_z^\alpha |\zeta|}}{2} \right\}, \tag{3.23}$$

where $R^p = (|\zeta|^2 + \frac{1}{\nu^\alpha} \rho^2)^{\frac{1}{2}}$ with $\alpha = e$ or h, ν^α being the anisotropy ratio. $\zeta = \zeta(z, z')$ denotes the vertical separation between observation and source points or *quasi-static* image points, as shown in Fig. 3.2. G_z^α is a half-line potential [4].

For instance, the acceleration of I_1 can be expressed as

Fig. 3.2 The direct radiation (black line) and reflection from lower and upper boundaries (blue line) when the source point and observation point are in the same layer

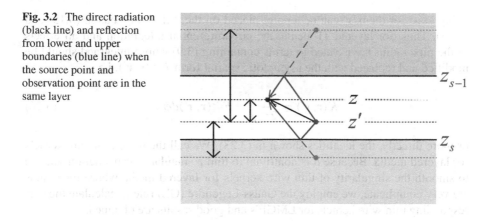

$$I_1 = S_0 \left\{ V_i^h - V_i^{h,\infty} \right\} + j\omega\mu_0\mu_t S_0 \left\{ \frac{1}{2jk_z^h}; 1, \overset{\leftarrow\alpha,\infty}{\Gamma_l}, \overset{\rightarrow\alpha,\infty}{\Gamma_l} \right\} \tag{3.24}$$

where the second part on the right hand side is defined by (3.25) with $\overset{\leftarrow\alpha,\infty}{\Gamma_l}$ and $\overset{\rightarrow\alpha,\infty}{\Gamma_l}$ representing the local reflection coefficient looking up and looking down. From the expression, we know that the first identity in (3.23) can be applied to obtain the asymptotic form of I_1. Accelerations of the remaining 13 independent integrals are appended.

$$S_m \left\{ \tilde{f}^\infty (k_t) ; \sigma, \overset{\leftarrow\alpha,\infty}{\Gamma_l}, \overset{\rightarrow\alpha,\infty}{\Gamma_l} \right\} =$$

$$\begin{cases}
\dfrac{1}{2\pi} \displaystyle\int_0^\infty \tilde{f}^\infty (k_t) \, \sigma \, e^{-jk_{zl}|z-z'|} J_m (k_t D) \, k_t dk_t \\[2mm]
\quad + \dfrac{1}{2\pi} \displaystyle\int_0^\infty \tilde{f}^\infty (k_t) \, \overset{\leftarrow\alpha,\infty}{\Gamma_l} \, e^{-jk_{zl}(z+z'-2z_l)} J_m (k_t D) \, k_t dk_t \\[2mm]
\quad + \dfrac{1}{2\pi} \displaystyle\int_0^\infty \tilde{f}^\infty (k_t) \, \overset{\rightarrow\alpha,\infty}{\Gamma_l} \, e^{-jk_{zl}(2z_{l+1}-z-z')} J_m (k_t D) \, k_t dk_t, \quad l = l' \\[2mm]
\dfrac{1}{2\pi} \displaystyle\int_0^\infty \tilde{f}^\infty (k_t) \, \sigma \, e^{-jk_{zl}|z-z'|} \left(1 + \overset{\rightarrow\alpha,\infty}{\Gamma_l} \right) J_m (k_t D) \, k_t dk_t, \quad l = l' + 1 \\[2mm]
\dfrac{1}{2\pi} \displaystyle\int_0^\infty \tilde{f}^\infty (k_t) \, \sigma \, e^{-jk_{zl}|z-z'|} \left(1 + \overset{\leftarrow\alpha,\infty}{\Gamma_l} \right) J_m (k_t D) \, k_t dk_t, \quad l = l' - 1
\end{cases} \tag{3.25}$$

3.2.3 Thin Wire Kernels for Layered TI Media

As discussed before, the spatial form of the extracted asymptotic terms or $F^\infty(\mathbf{r}, \mathbf{r}')$ is the singular parts of the LMGF's. When these terms are extracted, all the remaining spectral difference potential kernels either vanish or are slowly varying about the drill string axis; for them, $\mathbf{r}'$ can be a point placed on the drill string axis in (3.13) and (3.14). Thus, for layered TI media, the circumferential integration and all effects of the pipe radius are isolated to terms containing (3.9) whose integrand should be modified and replaced with the asymptotic spatial form $F^\infty(\mathbf{r}, \mathbf{r}')$, i.e.,

$$K(\mathbf{r}, \mathbf{r}') = \frac{1}{2\pi} \int_{-\pi}^{\pi} F^\infty(\mathbf{r}, \mathbf{r}') d\phi', \tag{3.26}$$

or more directly, the identities shown in (3.23). We call them the thin wire kernels for layered media. Because it is nontrivial to find a singularity cancellation scheme to smooth the singularity of thin wire kernels for layered media whose integrands are very complicate, we employ the Gauss-Legendre (GL) rule to calculate the corresponding thin wire kernels for LMGF's and good results are obtained.

3.2.4 Method of Moments

MoM transforms the governing Eq. (3.1) into a dense matrix equation to enable its solution by digital computers. Firstly, the drill string is subdivided into $(N + 1)$ linear segments with N interior nodes, as shown in Fig. 3.3. Since the unknown current vanishes at the endpoints of the thin wire, the unknown current $\mathbf{I}(\mathbf{r})$ is approximated as

$$\mathbf{I}(\mathbf{r}) = \sum_{n=1}^{N} I_n \Lambda_n(\mathbf{r}), \quad \mathbf{r} \in S, \tag{3.27}$$

where I_n is the unknown current coefficient on the nth interior node and Λ_n is a vector basis function used to represent the current on the wire. The triangle basis function and linear element are used in this paper. Higher order element and basis function can also be applied [6]. The triangle basis function is

$$\Lambda_n(\mathbf{r}) = \begin{cases} \dfrac{\rho_n^+}{h_n^+}, \ \mathbf{r} \in D_n^+; \\[2mm] \dfrac{\rho_n^-}{h_n^-}, \ \mathbf{r} \in D_n^-; \\[2mm] 0, \quad \text{elsewhere.} \end{cases} \tag{3.28}$$

where, as illustrated in Fig. 3.4, $D_n^\pm$ is the $\pm$ reference segment attached to the nth non-boundary node of a wire. The length of $D_n^\pm$ relative to the nth node is $h_n^\pm$, and $\rho_n^\pm$ is $(\pm)\times$ (the vector from the free node to $\mathbf{r}$). Inserting (3.3), (3.13), (3.14), and (3.27) into (3.1), and applying the Galerkin form of the MoM, we acquire

$$\{[Z_{mn}] + [Z_I]\}[I_n] = [V_m^i], \tag{3.29}$$

Fig. 3.3 The whole drill string is discretized into $(N + 1)$ linear segments with N interior nodes

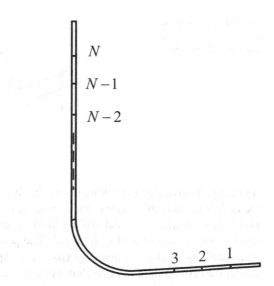

where $[Z_{mn}]$ represents the mutual impedance between nodes m and n, $[Z_I]$ is the matrix corresponding to the internal impedance of the drill string, and $[V_m^i]$ corresponds to the impressed voltage at the gap source. The evaluations of the $[Z_{mn}]$, $[Z_I]$ and $[V_m^i]$ are

$$Z_{mn} = \langle \Lambda_m; j\omega A(\Lambda_n) \rangle_{D_m} + \langle \Lambda_m; \nabla \Phi(\Lambda_n) \rangle_{D_m} \qquad (3.30)$$

$$Z_I = Z_i \langle \Lambda_m; \Lambda_n \rangle_{D_m} \qquad (3.31)$$

$$V_m = \langle \Lambda_m; \mathbf{E}^i(\mathbf{r}) \rangle_{D_m} \qquad (3.32)$$

where

$$\langle \mathbf{X}; \mathbf{Y} \rangle_{D_m} = \int_{D_m^+ + D_m^-} \mathbf{X} \cdot \mathbf{Y} \, dl \qquad (3.33)$$

and $A(\Lambda_n)$ and $\Phi(\Lambda_n)$ can be evaluated by (3.13) and (3.14).

The entries of the system matrix and vector can be evaluated by the Gaussian quadrature scheme and the singularity cancellation scheme. Once the current distribution on the drill string is solved out, it is easy to calculate the electromagnetic field at any position by using (3.3) and to obtain the voltage drop received by a surface receiver.

3.2.5 Modeling of the Drilling Fluid

The drill string is usually surrounded by drilling fluid which can be composed of water, oil, and gas. Different drilling fluid has different electric conductivity. It is

Fig. 3.4 Geometrical parameters associated with the nth node of wire

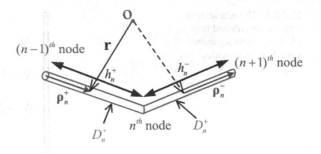

interesting to investigate the influence of the fluid to the EMT system. Assuming that the drill string is in the center of the borehole, Fig. 3.5 shows the cross section of the borehole with drill string and drilling fluid inside. The radius of the drill string is a and the outer radius of drilling fluid is b. The permeability and effective permittivity of the drilling fluid are denoted by $(\mu_0, \varepsilon_{fc})$. Applying the theory of VEP, we can model the drilling fluid as equivalent volume current

$$\mathbf{J}_v(\mathbf{r}) = j\omega(\varepsilon_{fc} - \varepsilon_b)\mathbf{E}(\mathbf{r}),\qquad(3.34)$$

where ε_b is the effective permittivity of the background formation. Then, the electric field inside the drilling fluid can be acquired using the current continuity equation and the quasi-static approximation since the working frequency of the EMT system is low. Using the equation of continuity, the charge density on the surface of the drill string is related to the surface current by

$$q_s = -\frac{1}{j\omega 2\pi a}\frac{dI(l)}{dl}\qquad(3.35)$$

For a perfectly conducting wire, the ρ component of the electric flux density D_ρ at the surface of the wire, is equal to q_s. The electric field on the surface of the uncoated wire then is approximated by

$$\mathbf{E}(\rho = a) = \hat{\rho}\frac{q_s}{\varepsilon_{fc}} = -\hat{\rho}\frac{1}{j\omega 2\pi a\varepsilon_{fc}}\frac{dI(l)}{dl}\qquad(3.36)$$

The tangential component of the electric field inside the fluid or borehole are considered to be negligible. To extrapolate from the value at the wire surface to points in the drilling fluid, we assume a $\frac{1}{\rho}$ radial variation which is appropriate for quasi-static approximation. Then this electric field in the fluid can be expressed as

$$\mathbf{E}(\rho) = -\hat{\rho}\frac{1}{j\omega 2\pi\rho\varepsilon_{fc}}\frac{dI(l)}{dl},\quad a \leq \rho \leq b\qquad(3.37)$$

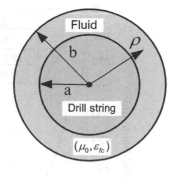

Fig. 3.5 The cross section of the borehole with drill string placed in the center and drilling fluid

which is inversely proportional to the radial distance ρ and is radial directed. The quasi-static property confines the equivalent charges to the surface of the drill string and the outer boundary of the drilling fluid. Combining (3.34) and (3.37), we can get the expression of the volumetric equivalent currents and equivalent charges using the axial current $\mathbf{I}(\mathbf{r})$ on the drill string. Hence, the modeling of the drilling fluid does not increase the number of unknowns. Compared to volume-surface integral equation (VSIE) approach for scattering from composite conducting-dielectric objects [15, 19], our approach only needs to solve the EFIE, not the combination of EFIE and the volume integral equation (VIE) in VSIE approach, since the volume equivalent current can be expressed by the axial current on the drill string. The only added calculations are

$$\mathbf{A}(\mathbf{r}) = \mu_0 \int_V \mathscr{G}^{\mathbf{A}}(\mathbf{r}, \mathbf{r}') \cdot \mathbf{J}_v \, dV', \tag{3.38}$$

$$\Phi(\mathbf{r}) = -\frac{1}{j\omega\varepsilon_0} \left[\int_V K^{\Phi}(\mathbf{r}, \mathbf{r}') \nabla'_S \cdot \mathbf{J}_v \, dV' + \int_V P_z(\mathbf{r}, \mathbf{r}') \hat{\mathbf{z}} \cdot \mathbf{J}_v \, dV' \right], \tag{3.39}$$

the volume integrals of the volume equivalent current and its divergence. Details of the volume integral can be referred to [12, 21], and no more details will be discussed.

3.3 Examples and Discussions

Based on the derivation, we implemented the modeling algorithms to solve the current distribution on the drill string and compute the excited electric field by the voltage/current gap source. One should note that the current distribution induced by one current gap source can be acquired by normalizing the current distribution induced by a voltage gap source. Both vertical and horizontal well in layered TI media are simulated to validate our method. Further simulations and results are discussed to illustrate the flexibility of our method to model various EMT systems. In all the simulations, the conductivity of air is taken as 10^{-6} S/m, and the relative permittivity and permeability of the layered formation are both one.

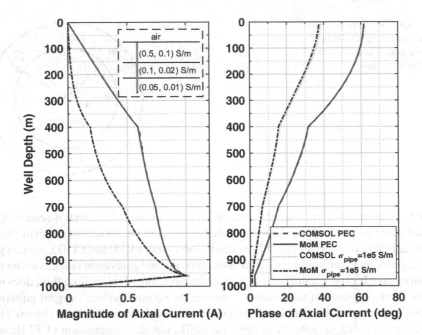

Fig. 3.6 Magnitude (left) and phase (right) of the axial current on the 1000-meter-long vertical well in a three-layer TI media

First, to validate the proposed method, a vertical well in layered TI media is studied. The formation has three layers excluding the half space filled with air on the top. The horizontal conductivity in each layer is, from top to bottom, 0.5 S/m, 0.1 S/m, 0.05 S/m, respectively. And the vertical conductivity in each layer is, from top to bottom, 0.1 S/m, 0.02 S/m, 0.01 S/m, respectively. The interfaces are at $z = 400$ m and $z = 700$ m. The 1000-meter-long vertical drill string has a radius of 12.7 cm. A $1-A$ current source is placed at depth $z = 960$ m and it works at 5 Hz. Figure 3.6 shows the magnitude and phase of the axial current distributions on the vertical drill string when the drill pipe is perfect electric conductor (PEC) and has a conductivity of 10^5 S/m. The results obtained by commercial FEM software COMSOL and the proposed method agree very well both in the magnitude and phase of the axial current. And from the results, we observe that the current decay faster inside layer with higher conductivity. This is reasonable because more current will leak into the more conductive formation resulting larger attenuation of the current on the drill string. Moreover, conductivity of the drill string also influences the attenuation rate of the current. As shown in the results, it is clear that the lower the conductivity of the drill string is, the faster the current along the well decays.

The cases studied up to now are limited to the vertical well. Hence, a more convincing case that has complicate well trajectory in layered TI media will be studied to demonstrate the efficiency and accuracy of our method. The 3D trajectory of the well and the layered TI media are shown in Fig. 3.7. The TI conductivity of layer i $(i = 1, 2, 3)$ is denoted by $(\sigma_{hi}, \sigma_{vi})$, where σ_{hi} and σ_{vi} denote the horizontal conductivity and the vertical conductivity, respectively. The interfaces are $z = 0$, 1000, 2000 m and $(\sigma_{h1}, \sigma_{v1}) = (0.5, 0.5)$ S/m, $(\sigma_{h2}, \sigma_{v2}) = (0.1, 0.02)$ S/m, $(\sigma_{h3}, \sigma_{v3}) = (0.05, 0.05)$ S/m. The conductivity of air is assumed to be 10^{-6} S/m. The 3D trajectory of the well composes of five parts among which three of them are straight lines and two of them are arcs. Three straight lines (black lines) shown in Fig. 3.7 can be expressed as (from top to bottom)

$$\begin{cases} x = 0 \\ y = 0 \qquad\qquad t \in [0, 1), \\ z = 1000 \times t \end{cases} \tag{3.40}$$

$$\begin{cases} x = 200 - 100\sqrt{2} + 500\sqrt{2} \times t \\ y = 0 \qquad\qquad\qquad\qquad t \in [0, 1), \\ z = 1000 + 100\sqrt{2} + 500\sqrt{2} \times t \end{cases} \tag{3.41}$$

and

$$\begin{cases} x = 200 + 500\sqrt{2} \\ y = -200 - 1000 \times t \quad t \in [0, 1], \\ z = -1000 - 700\sqrt{2} \end{cases} \tag{3.42}$$

respectively. Meanwhile, two arcs (green curves) connect the straight lines with a large radius of 200 m and the radians of the two arcs are 45 and 90°, as shown in Fig. 3.7. Hence, the whole well composes a vertical part, a deviated part, and a horizontal part. A $1-V$ voltage gap source is located at 100 m away from the head of the well. The total length of the well is 3471.24 m. It is assumed that the drill string is PEC.

In Fig. 3.8, the magnitude of axial current along the 3D well is shown in log scale when the EMT system employs the 1-V voltage source and works at 20 Hz. Measured depth represents the length along the trajectory of the drill string. As shown in the figure, the results obtained by COMSOL and the proposed method match very well and the current reaches its peak value at the position of the voltage gap source. Meanwhile, one can observe that even though the deviated part and horizontal part are in the same layer, the rate of decay of the current is different. Obviously, the reason for this difference is that the vertical conductivity is different from the horizontal conductivity. Meanwhile, from Fig. 3.8 one can observe that even though the deviated part (the middle black line in Fig. 3.7) and the horizontal part (the bottom black line in Fig. 3.7) of the well are in the same layer, the rates of the attenuation of the current magnitude on them are different. The attenuation rate of the current from 3300 to 2500 m (horizontal part) is smaller than that from 2200 to 1200 m (deviated part). Obviously, the reason for this discrepancy is that the vertical conductivity in that formation layer is one fifth of the horizontal conductivity. Comparing the computation cost of COMSOL and the proposed method as listed

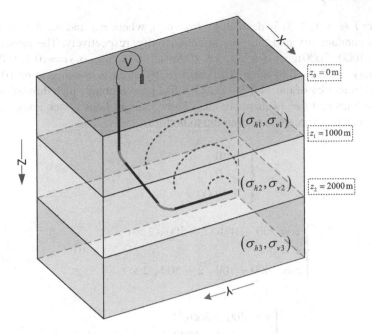

Fig. 3.7 A well with complicate 3D trajectory in three-layer TI media

Table 3.1 Computation cost for 3D case

	Memory cost	CPU time
COMSOL	497 GB	256 min
This method	6.5 MB	39 s

in Table 3.1, the proposed method is much more efficient to model the complicate well in layered TI media. Figure 3.9 shows two slices of the total electric field in the formation and air. The trajectory of the 3D well is also plotted as black curve to give a better illustration. While the transverse component electric field is continuous at the interface, the normal component is not. As a result, a clear pattern of layered structure can be observed from the discontinuous fields at the interfaces. More interestingly, because the drill string extends on a large scale in the second layer, the electric field in the second layer having high conductivity attenuates more slowly than the bottom layer having low conductivity. To study the behavior of this EMT system for the specific layered formation versus working frequency, Fig. 3.10 shows the reduction of the magnitude of the received signal by the surface antenna when the working frequency goes high. The surface antenna links the wellhead and a point in the $-y$ direction which is 100 m away from the wellhead. As the frequency is below 1 Hz, no significant change of the signal is observed. However, the signal decreases quickly and will go indistinguishable from the environmental noise when the frequency of the source is above 10 Hz.

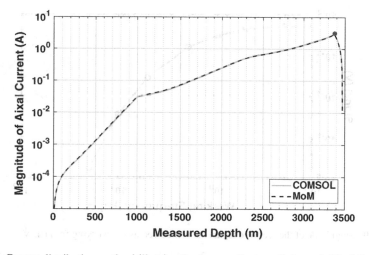

Fig. 3.8 Current distribution on the drill string for the complicate well shown in Fig. 3.7

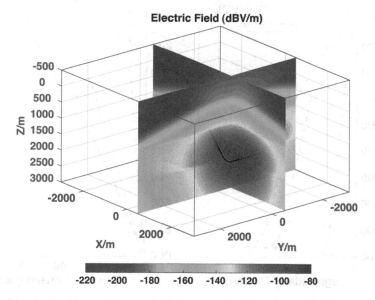

Fig. 3.9 Magnitude of the total electric field of two vertical slices for the EMT system shown in Fig. 3.7. Black curve represents the well trajectory

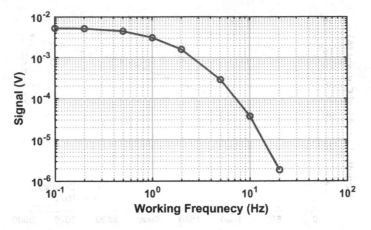

Fig. 3.10 Magnitude of the received signal at surface versus the working frequency

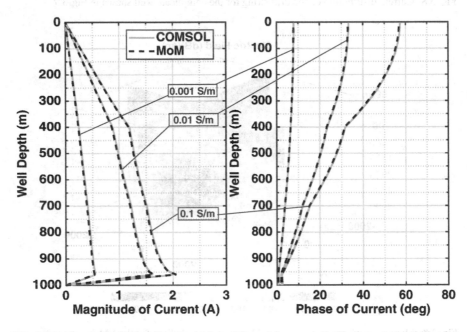

Fig. 3.11 Magnitude (left) and phase (right) of the axial current distributions on the drill string when drilling fluid is of different conductivity

The effect of the drilling fluid can be modeled by applying the equivalence principle. The various types of drilling fluid fall into a few broad categories, such as water-based mud and oil-based mud. The effects of drilling fluid to the EMT signal, under some circumstances, are not negligible and must be taken into consideration.

Inserted into the same layered TI media as depicted in Fig. 3.6, the 1000-meter-long vertical well is in the center of a borehole whose radius is 5.0 cm larger than the drill string, i.e., 17.7 cm. A $1-V$ voltage gap source is embedded at $z = 960$ depth and its frequency is 5 Hz. Figure 3.11 shows the axial current distributions on the drill string when the conductivity of the drilling fluid changes from 0.001 to 0.1 S/m, corresponding to changing from oil-based mud to water-based mud. The results obtained by our method closely match with the results obtained from COMSOL. As shown in the results of Fig. 3.11, compared to water-based mud, the oil-based mud reduces the strength and maximum of the current on the drill string substantially, thus leading to weaker signal received by the surface antenna. The oil-based mud raises the total resistivity of the formation and leads to a smaller axial current when constant voltage source is employed. We should point out that, if the gap source is a constant current source, the conclusion will be opposite, arriving at one that the current on the drill string attenuates faster in water-based mud compared to oil-based mud.

Appendix

The 14 independent integrals are expressed in the form of asymptotic singularity substraction.

$$I_1 = S_0 \left\{ V_i^h - V_i^{h,\infty} \right\} + j\omega\mu_0\mu_t S_0 \left\{ \frac{1}{2jk_z^h}; 1, \overleftarrow{T}^{h,\infty}, \overrightarrow{T}^{h,\infty} \right\}$$

$$I_2 = S_0 \left\{ I_v^e - I_v^{e,\infty} \right\} + j\omega\varepsilon_0\varepsilon_t S_0 \left\{ \frac{1}{2jk_z^e}; 1, -\overleftarrow{T}^{e,\infty}, -\overrightarrow{T}^{e,\infty} \right\}$$

$$I_3 = S_0 \left\{ \left(I_i^h - I_i^{h,\infty} \right) + \left(I_i^e - I_i^{e,\infty} \right) \right\} + S_0 \left\{ \frac{1}{2}; \text{sgn} \left| z - z' \right|, \overleftarrow{T}^{h,\infty}, -\overrightarrow{T}^{h,\infty} \right\}$$

$$+ S_0 \left\{ \frac{1}{2}; \text{sgn} \left| z - z' \right|, \overleftarrow{T}^{e,\infty}, -\overrightarrow{T}^{e,\infty} \right\}$$

$$I_4 = S_0 \left\{ \left(V_v^e - V_v^{e,\infty} \right) + \left(V_v^h - V_v^{h,\infty} \right) \right\} + S_0 \left\{ \frac{1}{2}; \text{sgn} \left| z - z' \right|, -\overleftarrow{T}^{h,\infty}, \overrightarrow{T}^{h,\infty} \right\}$$

$$+ S_0 \left\{ \frac{1}{2}; \text{sgn} \left| z - z' \right|, -\overleftarrow{T}^{e,\infty}, \overrightarrow{T}^{e,\infty} \right\}$$

$$I_5 = S_0 \left\{ \left(\frac{V_i^h}{k_\rho^2} - \frac{V_i^{h,\infty}}{(jk_z^h)^2/v^h} \right) - \left(\frac{V_i^e}{k_\rho^2} - \frac{V_i^{e,\infty}}{(jk_z^e)^2/v^e} \right) \right\}$$

$$+ j\omega\mu_0\mu_t v^h S_0 \left\{ \frac{1}{2(jk_z^h)^3}; 1, \overleftarrow{\Gamma}^{h,\infty}, \overrightarrow{\Gamma}^{h,\infty} \right\}$$

$$- \frac{1}{j\omega\varepsilon_0\varepsilon_t} v^e S_0 \left\{ \frac{1}{2jk_z^e}; 1, \overleftarrow{\Gamma}^{e,\infty}, \overrightarrow{\Gamma}^{e,\infty} \right\}$$

$$I_6 = S_0 \left\{ \left(\frac{V_v^h}{k_\rho^2} - \frac{V_v^{h,\infty}}{(jk_z^h)^2/v^h} \right) - \left(\frac{V_v^e}{k_\rho^2} - \frac{V_v^{e,\infty}}{(jk_z^e)^2/v^e} \right) \right\}$$

$$+ v^h S_0 \left\{ \frac{1}{2(jk_z^h)^2}; \operatorname{sgn}|z - z'|, -\overleftarrow{\Gamma}^{h,\infty}, \overrightarrow{\Gamma}^{h,\infty} \right\}$$

$$- v^e S_0 \left\{ \frac{1}{2(jk_z^e)^2}; \operatorname{sgn}|z - z'|, -\overleftarrow{\Gamma}^{e,\infty}, \overrightarrow{\Gamma}^{e,\infty} \right\}$$

$$I_7 = S_0 \left\{ \left(\frac{I_v^e}{k_\rho^2} - \frac{I_v^{e,\infty}}{(jk_z^e)^2/v^e} \right) - \left(\frac{I_v^h}{k_\rho^2} - \frac{I_v^{h,\infty}}{(jk_z^h)^2/v^h} \right) \right\}$$

$$+ j\omega\varepsilon_0\varepsilon_t v^e S_0 \left\{ \frac{1}{2(jk_z^e)^3}; 1, -\overleftarrow{\Gamma}^{e,\infty}, -\overrightarrow{\Gamma}^{e,\infty} \right\}$$

$$- \frac{1}{j\omega\mu_0\mu_t} v^h S_0 \left\{ \frac{1}{2jk_z^h}; 1, -\overleftarrow{\Gamma}^{h,\infty}, -\overrightarrow{\Gamma}^{h,\infty} \right\}$$

$$I_8 = S_0 \left\{ \left(\frac{I_i^e}{k_\rho^2} - \frac{I_i^{e,\infty}}{(jk_z^e)^2/v^e} \right) - \left(\frac{I_i^h}{k_\rho^2} - \frac{I_i^{h,\infty}}{(jk_z^h)^2/v^h} \right) \right\}$$

$$+ v^e S_0 \left\{ \frac{1}{2(jk_z^e)^2}; \operatorname{sgn}|z - z'|, \overleftarrow{\Gamma}^{e,\infty}, -\overrightarrow{\Gamma}^{e,\infty} \right\}$$

$$- v^h S_0 \left\{ \frac{1}{2(jk_z^h)^2}; \operatorname{sgn}|z - z'|, \overleftarrow{\Gamma}^{h,\infty}, -\overrightarrow{\Gamma}^{h,\infty} \right\}$$

$$I_9 = S_1 \left\{ \left(\frac{I_i^e - I_i^{e,\infty}}{k_\rho} \right) - \left(\frac{I_i^h - I_i^{h,\infty}}{k_\rho} \right) \right\} + S_1 \left\{ \frac{1}{2k_\rho}; \text{sgn} \, |z - z'|, \overleftarrow{T}^{e,\infty}, -\overrightarrow{T}^{e,\infty} \right\}$$

$$- S_1 \left\{ \frac{1}{2k_\rho}; \text{sgn} \, |z - z'|, \overleftarrow{T}^{h,\infty}, -\overrightarrow{T}^{h,\infty} \right\}$$

$$I_{10} = S_1 \left\{ \left(\frac{V_v^h - V_v^{h,\infty}}{k_\rho} \right) - \left(\frac{V_v^e - V_v^{e,\infty}}{k_\rho} \right) \right\} + S_1 \left\{ \frac{1}{2k_\rho}; \text{sgn} \, |z - z'|, -\overleftarrow{T}^{h,\infty}, \overrightarrow{T}^{h,\infty} \right\}$$

$$- S_1 \left\{ \frac{1}{2k_\rho}; \text{sgn} \, |z - z'|, -\overleftarrow{T}^{e,\infty}, \overrightarrow{T}^{e,\infty} \right\}$$

$$I_{11} = S_1 \left\{ k_\rho \left[V_i^h - V_i^{h,\infty} \right] \right\} + j\omega\mu_0\mu_t S_1 \left\{ \frac{k_\rho}{2jk_z^h}; 1, \overleftarrow{T}^{h,\infty}, \overrightarrow{T}^{h,\infty} \right\}$$

$$I_{12} = S_1 \left\{ k_\rho \left[I_v^e - I_v^{e,\infty} \right] \right\} + j\omega\varepsilon_0\varepsilon_t S_1 \left\{ \frac{k_\rho}{2jk_z^e}; 1, -\overleftarrow{T}^{e,\infty}, -\overrightarrow{T}^{e,\infty} \right\}$$

$$I_{13} = S_2 \left\{ \left(I_i^e - I_i^{e,\infty} \right) - \left(I_i^h - I_i^{h,\infty} \right) \right\} + S_2 \left\{ \frac{1}{2}; \text{sgn} \, |z - z'|, \overleftarrow{T}^{e,\infty}, -\overrightarrow{T}^{e,\infty} \right\}$$

$$- S_2 \left\{ \frac{1}{2}; \text{sgn} \, |z - z'|, \overleftarrow{T}^{h,\infty}, -\overrightarrow{T}^{h,\infty} \right\}$$

$$I_{14} = S_2 \left\{ \left(V_v^h - V_v^{h,\infty} \right) - \left(V_v^e - V_v^{e,\infty} \right) \right\} + S_2 \left\{ \frac{1}{2}; \text{sgn} \, |z - z'|, -\overleftarrow{T}^{h,\infty}, \overrightarrow{T}^{h,\infty} \right\}$$

$$- S_2 \left\{ \frac{1}{2}; \text{sgn} \, |z - z'|, -\overleftarrow{T}^{e,\infty}, \overrightarrow{T}^{e,\infty} \right\}$$

References

1. Aksun, M., Dural, G.: Clarification of issues on the closed-form green's functions in stratified media. IEEE Trans. Antennas Propag. **53**(11), 3644–3653 (2005)
2. Alparslan, A., Aksun, M.I., Michalski, K.A.: Closed-form green's functions in planar layered media for all ranges and materials. IEEE Trans. Microwave Theory Tech. **58**(3), 602–613 (2010)
3. Bhagwan, J., Trofimenkoff, F.N.: Electric drill stem telemetry. IEEE Trans. Geosci. Remote Sens. **GE-20**(2), 193–197 (1982)
4. Celepcikay, F.T., Wilton, D.R., Jackson, D.R., Paulotto, S., Johnson, W.A.: Efficient evaluation of half-line source potentials and their derivatives. IEEE Trans. Antennas Propag. **60**(12), 5834–5842 (2012)

5. Champagne, N.J.: A three-dimensional method of moments formulation for material bodies in a planar multi-layered medium. Ph.D. dissertation, University of Houston (1996)
6. Champagne, N.J., Wilton, D.R., Rockyway, J.D.: The analysis of thin wires using higher order elements and basis functions. IEEE Trans. Antennas Propag. **54**(12), 3815–3821 (2006)
7. DeGauque, P., Grudzinski, R., et al.: Propagation of electromagnetic waves along a drillstring of finite conductivity. SPE Drill Eng. **2**(02), 127–134 (1987)
8. Dural, G., Aksun, M.I.: Closed-form green's functions for general sources and stratified media. IEEE Trans. Microwav. Theory Tech. **43**(7), 1545–1552 (1995)
9. Graglia, R.D.: On the numerical integration of the linear shape functions times the 3-D green's function or its gradient on a plane triangle. IEEE Trans. Antennas Propag. **41**(10), 1448–1455 (1993)
10. Hill, D.A., Wait, J.R.: Electromagnetic basis of drill-rod telemetry. Electron. Lett. **14**(17), 532–533 (1978)
11. Johnston, R.H., Trofimenkoff, F., Haslett, J.W.: Resistivity response of a homogeneous earth with a finite-length contained vertical conductor. IEEE Trans. Geosci. Remote Sens. **4**, 414–421 (1987)
12. Khayat, M.A.: Numerical modeling of thin materials in electromagnetic scattering problems. Ph.D. thesis, University of Houston (2003)
13. Li, D.: Efficient computation of layered medium green's function and its application in geophysics. Ph.D. dissertation, University of Houston (2016)
14. Li, D., Wilton, D.R., Jackson, D.R.: Recent advances in evaluating green's functions for multi-layered media and half-space problems. In: Computing and Electromagnetics International Workshop (CEM), 2017, pp. 1–2. IEEE (2017)
15. Lu, C.C., Chew, W.C.: A coupled surface-volume integral equation approach for the calculation of electromagnetic scattering from composite metallic and material targets. IEEE Trans. Antennas Propag. **48**(12), 1866–1868 (2000)
16. Michalski, K.A.: Extrapolation methods for sommerfeld integral tails. IEEE Trans. Antennas Propag. **46**(10), 1405–1418 (1998)
17. Michalski, K.A., Mosig, J.R.: Multilayered media green's functions in integral equation formulations. IEEE Trans. Antennas Propag. **45**(3), 508–519 (1997)
18. Mosig, J.: The weighted averages algorithm revisited. IEEE Trans. Antennas Propag. **60**(4), 2011–2018 (2012)
19. Nie, X.-C., Yuan, N., Li, L.-W., Gan, Y.-B., Yeo, T.S.: A fast volume-surface integral equation solver for scattering from composite conducting-dielectric objects. IEEE Trans. Antennas Propag. **53**(2), 818–824 (2005)
20. Peterson, A.F., Ray, S.L., Mittra, R.: Computational methods for electromagnetics. IEEE Press, New York (1998)
21. Schaubert, D., Wilton, D., Glisson, A.: A tetrahedral modeling method for electromagnetic scattering by arbitrarily shaped inhomogeneous dielectric bodies. IEEE Trans. Antennas Propag. **32**(1), 77–85 (1984)
22. Schnitger, J., Macpherson, J.D., et al.: Signal attenuation for electromagnetic telemetry systems. SPE/IADC Drilling Conference and Exhibition. Society of Petroleum Engineers (2009)
23. Simsek, E., Liu, Q.H., Wei, B.: Singularity subtraction for evaluation of green's functions for multilayer media. IEEE Trans. Microwav. Theory Tech. **54**(1), 216–225 (2006)
24. Trofimenkoff, F.N., Segal, M., Klassen, A., Haslett, J.W.: Characterization of EM downhole-to-surface communication links. IEEE Trans. Geosci. Remote Sens. **38**(6), 2539–2548 (2000)
25. Wait, J.R.: The effect of a buried conductor on the subsurface fields for line source excitation. Radio Sci. **7**(5), 587–591 (1972)
26. Wei, Y.: Propagation of electromagnetic signal along a metal well in an inhomogeneous medium. Ph.D. thesis, Norwegian University of Science and Technology (2013)
27. Weiss, C.J.: Finite-element analysis for model parameters distributed on a hierarchy of geometric simplices. Geophysics **82**(4), E155–E167 (2017)
28. Wilton, D.R., Champagne, N.J.: Evaluation and integration of the thin wire kernel. IEEE Trans. Antennas Propag. **54**(4), 1200–1206 (2006)

29. Wind, J., Weisbeck, D., Culen, M., et al.: Successful integration of electromagnetic MWD-LWD technology extends UBD operation envelope into severely depleted fields. In: SPE/IADC Drilling Conference. Society of Petroleum Engineers (2005)
30. Xia, M.Y., Chen, Z.Y.: Attenuation predictions at extremely low frequencies for measurement-while-drilling electromagnetic telemetry system. IEEE Trans. Geosci. Remote Sens. 31(6), 1222–1228 (1993)
31. Yang, J., Liu, Y., Wu, X.: 3-D DC resistivity modelling with arbitrary long electrode sources using finite element method on unstructured grids. Geophys. J. Int. 211(2), 1162–1176 (2017)
32. Yang, W., Torres-Verdn, C., Hou, J., Zhang, Z.I.: 1D subsurface electromagnetic fields excited by energized steel casing. Geophysics 74(4), E159–E180 (2009)
33. Zeng, S., Li, D., Wilton, D.R., Chen, J.: Fast and accurate simulation of electromagnetic telemetry in deviated and horizontal drilling. J. Petroleum Sci. Eng. 166, 242–248 (2018)

Chapter 4
Applications

4.1 Single Well Borehole Wireless Communication

In practice, the deployment of EM-MWD services and equipment for drilling a
planned total depth well of 10,000 ft, if lacking the benefits of planning based on
rigorous EMT simulations, may be described as follows. Many EM-MWD systems
feature variable and selectable carrier frequencies and transmitter power levels; generally, the carrier frequencies range from 2 to 12 Hz, the power levels 10–30, or as high
as 50 W. With little to no knowledge of formation resistivities and their influences
on EM signal attenuation, the directional drilling/MWD service would likely select
the higher end of the carrier frequency, and the lower end of the power spectrum, for
drilling the first few thousand feet. As drilling depth increases, the service company
would decrease the carrier frequency and increase the transmitting power level, rea-
soning that signal attenuation increases with increasing depth, and warrant the lower
frequency/higher power combination. By 10,000 ft, the EM-MWD system may be
operating at only 2 Hz but drawing 30–50 W. If the rate of penetration is relatively
low, the service provider may face a situation of rapid battery power depletion, caus-
ing the rig contractor/operator to incur expensive non-productive time in the form of
a trip out of the borehole to replace depleted batteries. An accurate and efficient EMT
simulation algorithm and software package, if available, will help to adaptively opti-
mize the deployment of an EMT system in different formation and drilling depths,
extend battery life, reduce non-productive time, and ultimately, increase profitability. However, since 3D modeling EMT in directional and horizontal drilling using
conventional numerical techniques is too expensive, in field practices a horizontal
or deviated well is approximated by a vertical well with the same measured depth.
Then quasi-analytical methods or 2.5D numerical schemes will be used to model the
simplified vertical well. Apparently this strategy will greatly speed up the simulation,
but at the expense of accuracy.

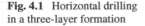

Fig. 4.1 Horizontal drilling
in a three-layer formation

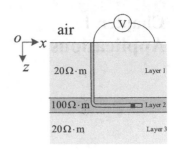

Figure 4.1 shows a directional drill string in a three-layer formation. The vertical portion of the drill string is 5000 ft long, while the horizontal portion is 3000 ft long. Thicknesses of Layers 1 and Layer 2 (the target layer) are 4900 ft and 200 ft, respectively. A 1 V gap source is located on the drill string 200 ft away from the drill bit. The resistivities of Layers 1–3 are 20 Ω m, 100 Ω m, and 20 Ω m, respectively. The operating frequency is 5 Hz and the radius of the drill string is 5 in. The surface receiver is deployed to detect the voltage drop between the wellhead and a point on the surface 300 ft away and in the same direction as the horizontal part of the drill string. The current distributions in both the simplified vertical drilling and the actual horizontal drilling are shown in Fig. 4.2, from which we can observe substantial differences between these two scenarios. Actually, the numerical result of received signal is 16.5 mV for the realistic horizontal drilling case, while it is 13.0 mV in simplified vertical drilling case. This example suggests that it is necessary to study the performance of an EMT system for a directional and horizontal drilling job using rigorous 3D numerical schemes, such as the IEM discussed in Chap. 3.

Figures 4.3 and 4.4 show current distributions and received signal strengths versus operating frequencies, respectively. The simulation results suggest that the received signal level decreases as the frequency increases. When EM telemetry is used for very deep measurements, the working frequency should be much lower if only surface antenna or voltage meter is deployed to detect the transmitted signal. One possible way to enhance the amplitude of the received signal is using the steel casing of adjacent completed well as one electrode [5]. Since the steel casing of adjacent well is much closer to the source, the magnitude of received signal can be enhanced by more than 10 times compared to conventional surface antenna. Relay of transceivers (i.e. repeaters) can also be applied to EM telemetry to obtain stronger signal [2].

Figures 4.5 and 4.6 show current distributions and received signal strengths versus resistivity of the layer containing the horizontal portion of the drill string (Layer 2 in Fig. 4.1), respectively. The operating frequency is fixed at 5 Hz. This example suggests that the conductivity of the target layer will affect the receivability of an EM telemetry system in a non-monotonic fashion. On the other hand, Fig. 4.6 demonstrate that the EMT signal will be suppressed more heavily as the formation between the drill bit and surface becomes more conductive.

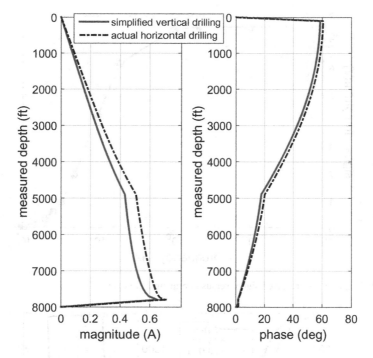

Fig. 4.2 Current distributions of simplified vertical drilling and of actual horizontal drilling

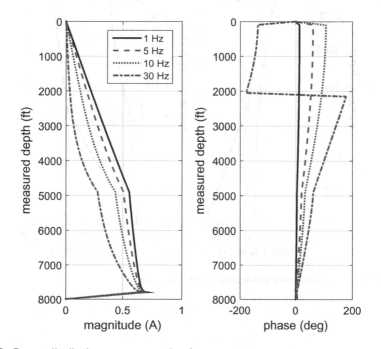

Fig. 4.3 Current distributions versus operating frequency

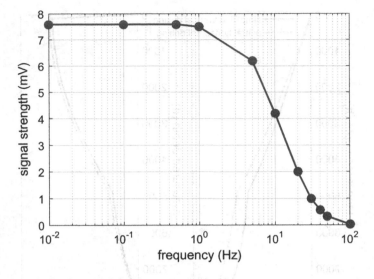

Fig. 4.4 Magnitude of received signal versus operating frequency

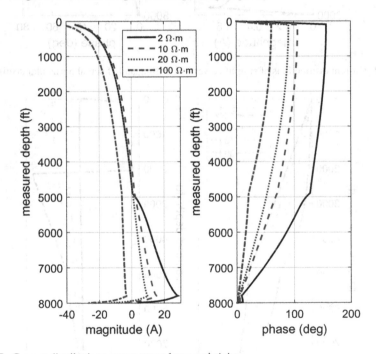

Fig. 4.5 Current distributions versus target layer resistivity

Fig. 4.6 Magnitude of received signal versus target layer resistivity

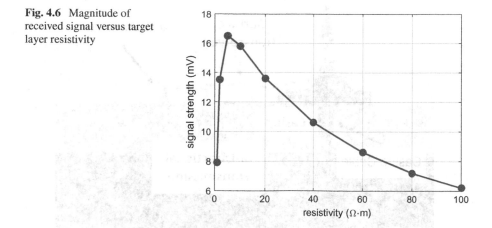

4.2 Long Range EMT Using Casing as Antenna

An innovative EMT system [1, 4] has been proposed and implemented for receiving stronger signals in conductive media and for a higher data transmission rate. In a traditional EMT, the surface transceiver connects one of its two terminals to the blowout preventer (BOP), and the other terminal to the earth antenna as a metal stake driven into the ground with a certain distance away from the rig. In other words, it measures the voltage drop between BOP and earth antenna. The new EMT system measures the voltage drop along the whole casing string: one terminal of the surface transceiver is still connected to BOP, and the other terminal is connected to the bottom of the casing, with a long cable externally attached to the whole casing string. The downhole cable is electrically connected to the bottom of the casing. Except that point the whole cable is insulated from the casing string all the way from top to bottom. Another deployment of the new EMT system is sort of a combination of the traditional EMT and the new system measuring along the casing string: the downhole cable is still connected to the bottom of the casing, but the surface cable is connected to the earth antenna instead of the BOP. Either way, the new EM telemetry measures the voltage drop between the casing bottom and a point on the surface. The schematics of different EMT systems are shown in Fig. 4.7.

Figure 4.8 shows the deployment of the downhole coaxial cable attached to the casing string: the left panel shows one end of the cable electrically connected to the casing bottom by a splice sub; the center panel shows the long coaxial cable goes all the way along the casing string, it is fixed to but electrically isolated from the casing by clamps at some intervals; and the right panel shows that at the surface the cable exists the well head and will be connected to one terminal of the surface transceiver. The deployment of this telemetry system on the surface is shown in Fig. 4.9. From which we can see how spool, sheave, and clamps are used to attach the coaxial cable to casing pipe.

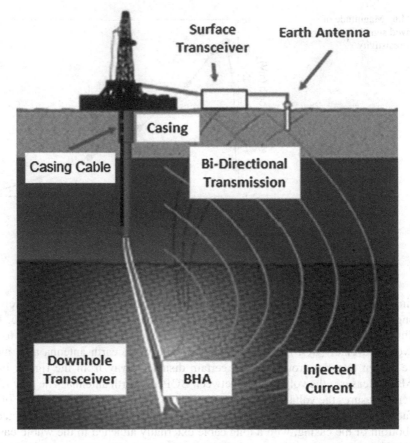

Fig. 4.7 Different EMT systems: the traditional EMT measures the potential difference between BOP and the earth antenna; the new EMT measures the voltage drop along the casing string, or the potential difference between the bottom of the casing and the earth antenna

The new EMT system with casing antenna and the traditional EMT are modeled and compared based on the rigorous numerical simulations. Since the new EMT measures the voltage drop along the casing string, the resistance of the casing pipe becomes an importance factor for this new system. Figure 4.10 compares the signal strengths of the two telemetry systems w.r.t. the total resistance of the whole casing string. In this case we assume the drilling depth is 10,000 ft and EMT systems work at 5 Hz, the underground formation is set as a homogeneous medium with resistivity as 10 Ω m. From this figure we can see that the signal strength of a traditional EMT decreases as the casing resistance increase, while the new EMT system with casing cable has the opposite trend (numerical modeling suggests that the signal of the new system will eventually drop if the casing resistance keeps increase to a fairly large number, which cannot be seen in the range of casing resistance used in this figure). Figure 4.10 shows that for this specific case, the new telemetry will provide stronger

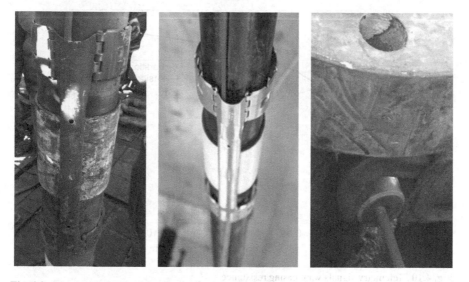

Fig. 4.8 The downhole coaxial cable deployment. Left: one end of the cable is electrically connected to the bottom of the casing string; Center: the cable is externally attached to the casing by clamps at some intervals; Right: the cable exists the well head

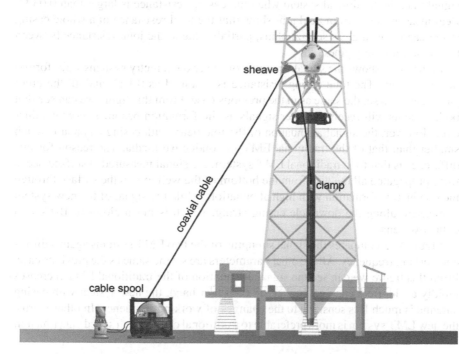

Fig. 4.9 The deployment of the telemetry system on the surface

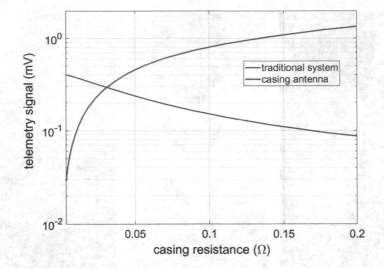

Fig. 4.10 Telemetry signals w.r.t. casing resistance

signal than the traditional system when the casing resistance is larger than 0.03 Ω. Real measurements from field jobs show that the total resistance of a whole casing string can be as large as several ohms, partially due to the joint resistance between two adjacent casing pipes.

Figure 4.11 shows the signal strengths of the two telemetry systems w.r.t. formation resistivity. The total casing resistance is assumed as 0.1 Ω, and all the other parameters are set the same as in the previous case. From this figure we can see that both systems will receive weaker signals as the formation becomes more conductive. However, the signal attenuation of the telemetry with casing antenna is much smaller than that of the traditional EMT in conductive media. The reason for this difference is that for a traditional EMT system, the signal measured at surface needs to be propagated all the way from the bottom of the wellbore to the surface through the conductive formation with high attenuation, while the signal of the new system is measure along the downhole casing string, which is much closer to BHA in a collective sense.

Figure 4.12 compares the signal strengths of the two EMT systems against different working frequency. All the other parameters are set the same as the previous case. From this figure we can see the signal attenuation of the traditional EMT increases rapidly as the frequency get higher, on the other hand, the new system with casing antenna is much less sensitive to the changes of working frequency. In other words, the new EMT system is more preferable to traditional one for high speed data transfer.

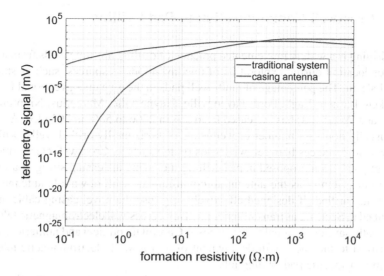

Fig. 4.11 Telemetry signals w.r.t. formation resistivity

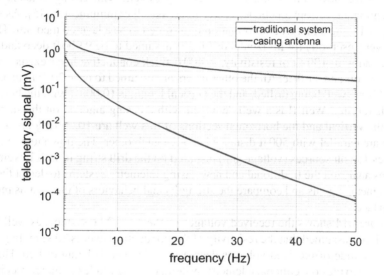

Fig. 4.12 Telemetry signals w.r.t. working frequency

4.3   Crosswell Communication in Pad Drilling

Pad drilling is a widely implemented practice of drilling multiple wells from a single surface location during exploration of unconventional resources, such as shale gas and shale oil. The percentage of multi-well pads in the nine unconventional US plays (Bakken, Barnett, Eagle Ford, Fayetteville, Haynesville, Marcellus, Niobrara, Permian, and Woodford) has increased from 5% in 2006 to 58% in 2013 [3]. A drilling pad usually houses a number of wellheads of horizontally drilled wells, with relatively short distances between wellheads on the surface. Crosswell electromagnetic telemetry has been proposed in pad drilling, i.e. using the casing string of a finished well in the vicinity as the antenna to communicate with downhole antenna in the well being drilled. If designed effectively (e.g. use the novel casing cable antenna described in Sect. 4.2 as transmitter or receiver), crosswell electromagnetic telemetry can greatly save operational time and costs. Careful theoretical and numerical studies are critical to the successful deployment of a crosswell electromagnetic telemetry system in a specific pad drilling job.

A series of numerical simulations were conducted to demonstrate the advantages of the novel electromagnetic telemetry system with casing antenna versus conventional electromagnetic telemetry in crosswell communication [5]. As shown in Fig. 4.13, the underground formation is assumed to be a layered medium. The target layer (reservoir) for horizontal drilling is assumed to be 9,900 ft deep and 200 ft thick, and with 100 Ω m resistivity. A 400 ft thick conductive layer exists between the reservoir and surface. All the other layers are assumed to have 20 Ω m resistivity. Well I is the well being drilled, and the vertical length is 10,000 ft with the horizontal length varying. Well II is a well deployed with a casing antenna on the same pad, both the vertical and the horizontal sections of this well are 10,000 ft long. The two wells are parallel with 500 ft distance between each other. The downhole telemetry utilizes 18 volt sources (voltage gap) installed on the drill string in well I. Numerical studies analyzed the traditional and new casing telemetry systems to detect the electromagnetic signals and compare the strengths and behaviors of the signals received on surface.

Figure 4.14 shows the received voltages of the two EMT systems as well II was drilled. It is assumed that the resistivity of the conductive layer is 20 Ω m. In general, the magnitude of received voltages decrease as the well is being drilled. However compared to the conventional telemetry system, the magnitude of voltage received by casing antenna is much larger. Hence, when the conventional telemetry system would not be able to provide distinguishable signals from noise as the well is being drilled further, the casing telemetry system presents a strong alternative without significant cost, especially in pad drilling.

In Fig. 4.15 we compared the received signals of the two electromagnetic telemetry systems versus the resistivity changes of the conductive layer between the reservoir and the surface. Numerical results show that the signal received by the conventional electromagnetic telemetry becomes weaker as this layer becomes more conductive, and this is because higher conductivity will attenuate electromagnetic signals more

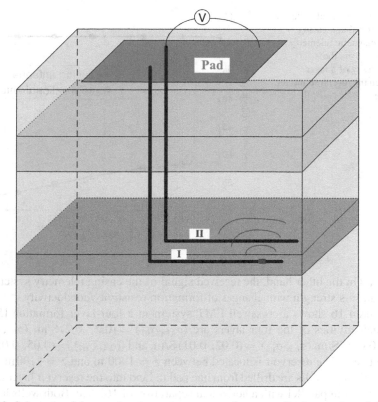

Fig. 4.13 Schematic of the novel casing antenna system for crosswell telemetry in pad drilling. Well I is being drilled, and the novel casing antenna is deployed in well II

Fig. 4.14 Received voltages of traditional and casing electromagnetic telemetry systems versus the measured depth of well I

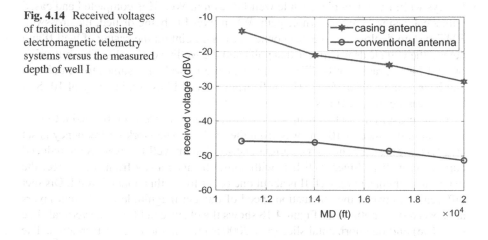

Fig. 4.15 Received voltages of traditional and casing electromagnetic telemetry systems versus the conductivity of a layer between the reservoir and the surface

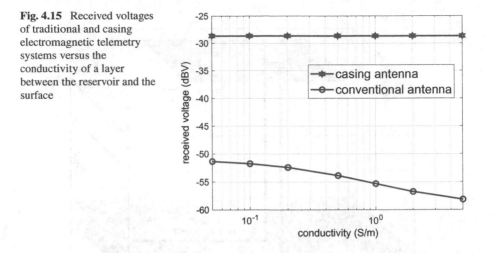

heavily. On the other hand, the received signal of the casing telemetry system stays maintains its strength with changes of formation resistivity/conductivity.

Figure 4.16 shows a crosswell EMT system in a four-layer formation [5]. The TI conductivities of the four layers are $(\sigma_{h1}, \sigma_{v1}) = (0.2,\ 0.2)$ S/m, $(\sigma_{h2}, \sigma_{v2}) = (0.1,\ 0.025)$ S/m, $(\sigma_{h3}, \sigma_{v3}) = (0.02,\ 0.01)$ S/m, and $(\sigma_{h4}, \sigma_{v4}) = (0.05,\ 0.05)$ S/m, respectively. The reservoir is located between $z = 1900$ m and $z = 2100$ m planes. Two horizontal wells are drilled from one pad to land into the reservoir layer at depth $z = 2000$ m in parallel with a horizontal separation of 167.6 m. Both wells last until reaching a horizontal length of 2000 m. The distance between the vertical parts of two wells is 30 m. The curvatures of the curved part of both wells are 300 m though well II stays within $y = 0$ m plane while well I does not. Well II is completed and cased with steel casing along the whole path. When well I is being drilled, the crosswell EMT system deploys a cable that connects to one point on the casing of well II and detects the voltage drop between that point and the wellhead of well I. The drill string of well I has a radius of 0.127 m, and the casing of well II assumes to be uniform and have a radius of 0.254 m. The drill string of well I has conductivity of 10^6 S/m and the casing of well II has conductivity of 10^5 S/m.

In the crosswell EMT system, the 1 V voltage gap source is embedded into well I and has a distance of 100 m away from the drill bit. The working frequency is set as 20 Hz. Figure 4.17 shows the current distribution on well I as well as the induced current on well II. Since well II is with some distance away from the source, the magnitude of current on well II is about one order lower than that of well I. Distinct differences between the attenuation speed of current magnitude in different layers are observed for both wells. Figure 4.18 shows the electric fields on one vertical slice ($y = 0$ m) and one horizontal slice ($z = 2000$ m) inside the layered formation. The trajectory of two horizontal wells are also plotted to give a better illustration. The electric field is strongest near the voltage source and attenuates along the well further away from the source. Interfaces between the layers can be discerned easily from the

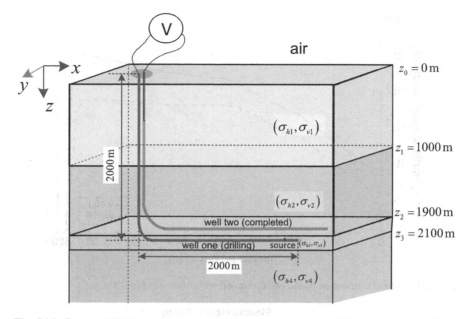

Fig. 4.16 Crosswell EM telemetry system. Well one is being drilled while well two is completed and cased

pattern of the field. To demonstrate the advantage of crosswell EMT system for pad drilling, the behaviors of the strength of the received signal for conventional EMT system and and of crosswell EMT system are compared in Fig. 4.19. For conventional EMT system, the surface antenna detects the voltage drop between the wellhead of well I and one surface point 100 m away from the wellhead on the surface. While for crosswell EMT system, the deployed cable inside well II connects to the casing at depth of 600 m. When well I is being drilled from the trajectory having horizontal length of 500 m to one having horizontal length of 2000 m, the detected signal of crosswell EMT system is about one order of magnitude stronger than that of the conventional EMT system.

This novel electromagnetic telemetry system has been used in several North American regions. One deployment was conducted in Zavala County, Texas. The depth of casing shoe is 7300 ft, and the casing OD is 7 in. Figure 4.20 shows a comparison between the traditional electromagnetic telemetry system and the new casing telemetry system. Approximately 25-fold (28 dB) signal gain was achieved by the electromagnetic casing telemetry system.

In the second field test, the novel casing cable was placed in in an offset well 360 m away from the drilling well. The OD of the casing is 7 in, and the depth of offset well casing shoe is 3170 m. The total length of the casing cable is 3105 m. The new electromagnetic casing telemetry system was employed to measure signals starting from 3265 m MD (3134 m TVD) and finishing at 6212 m MD (3110 m TVD) with oil-based mud. The signal is 2.14 mV at the beginning and 0.2 mV at the end.

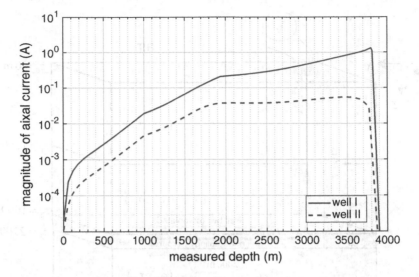

Fig. 4.17 Current distribution on the two horizontal wells shown in Fig. 4.16

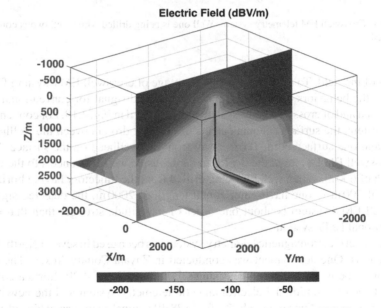

Fig. 4.18 Magnitude of the total electric field of two slides for the crosswell EMT system shown in Fig. 4.16

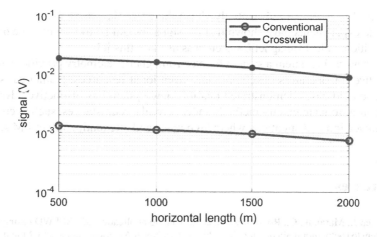

Fig. 4.19 Comparison of received signals by conventional and crosswell EMT systems

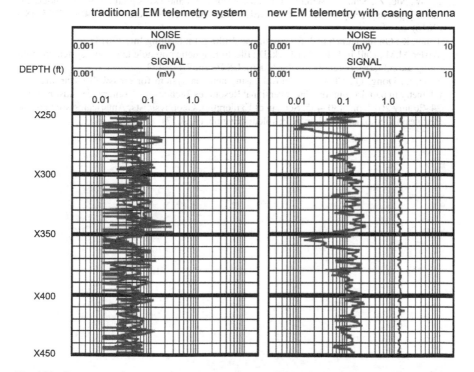

Fig. 4.20 Comparison between traditional electromagnetic telemetry and the new telemetry system with casing antenna. The left panel shows the strengths of signal and noise of the traditional electromagnetic telemetry, and the right panel shows the signal and noise of the new electromagnetic telemetry system with casing antenna

The signal was always substantially above the noise threshold. Simulation results and field experiments have shown that the signal would be too small to be detected if a traditional electromagnetic system was used on this job.

The new casing antenna system for electromagnetic telemetry provides a reliable alternative for operators developing unconventional resources. Numerical simulations and field cases demonstrated that this new system has distinctive advantages over conventional telemetry methods under several circumstances, such as crosswell communication and situations where high noise and attenuation affect EM signals.

References

1. Hunter, J., Maranuk, C., Rodriguez, A., et al.: Unique application of EM LWD casing antenna system to rocky mountain drilling. In: SPE Western North American and Rocky Mountain Joint Meeting. Society of Petroleum Engineers (2014)
2. Li, W., Nie, Z., Sun, X.: Wireless transmission of mwd and lwd signal based on guidance of metal pipes and relay of transceivers. IEEE Trans. Geosci. Remote Sens. **54**(8), 4855–4866 (2016)
3. Thuot, K.: On the launch pad: the rise of pad drilling. In: Drilling Info (2014)
4. Wisler, M.M., Hall Jr, H.E., Weisbeck, D.: Electromagnetic borehole telemetry system incorporating a conductive borehole tubular (2006). US Patent 7,145,473
5. Zeng, S., Dong, Q., Chen, J.: A novel casing antenna system for crosswell electromagnetic telemetry in pad drilling. In: Unconventional Resources Technology Conference, Austin, Texas, 24–26 July 2017, pp. 430–435. Society of Exploration Geophysicists, American Association of Petroleum Geologists, Society of Petroleum Engineers (2017)

Index

© The Author(s), under exclusive license to Springer Nature Switzerland AG 2019
J. Chen et al., *Borehole Electromagnetic Telemetry System*,
SpringerBriefs in Petroleum Geoscience & Engineering,
https://doi.org/10.1007/978-3-030-21537-8